The Anywhere Library
A Primer for the Mobile Web

by Courtney Greene, Missy Roser, and Elizabeth Ruane

Association of College and Research Libraries
A division of the American Library Association
Chicago, 2010

Table of Contents

Introduction

When you combine three public-services librarians, an equal number of smartphones, plenty of discussion, and lots of coffee, you might just end up with a book. As we logged hours on the reference desk, taught research classes, talked with students and student workers, and observed trends across our campus, it became clear to us that our community had reached a tipping point in terms of adoption of mobile devices. At this crossroads, we boldly committed to action—and then realized the enormity of everything we didn't know. As librarians, we recognize an information need when we see one, so we spent the next several months educating ourselves, making lots of mistakes, and learning from them. Our guiding philosophy has been "Make it happen, then make it better."

You, our reader, may also feel that your institution is ready to take the plunge. Numerous studies show the usage of mobile devices has moved beyond early adopters into the mainstream. A Deloitte telecom forecast pointed out that within one year, smartphone sales were expected to exceed sales of all other PCs combined, including netbooks, laptops, and desktops (Ruffolo 2010, 4). Mobile computing is rapidly becoming an important part of our everyday lives, and this is just one of many areas in which libraries can contribute to IT leadership on campus. With a user-centered, practical emphasis geared to the non-technical librarian, this book approaches the creation of a mobile-optimized library website as a process rather than simply a product. Whatever your background, regardless of your level of expertise with technology, whether or not you currently own a smartphone (although they're lots of fun, so you might want to give it a try!), you too can pull the trigger and launch a mobile site for your library.

Chapter One begins with an overview of who's using mobile devices on campus today, the implications for your library and staff, and how a needs assessment can ground your plan for mobile services in your community's needs and make it strategic. Chapter Two guides you through the selection and decision-making process for what will be included on your mobile website, building from your library services that are already heavily used. While this book focuses on mobile-optimized sites, we'll also discuss device-specific applications and how you might want to incorporate them. Chapter Three builds on the opportunities you've identified and the priorities you've set, transforming them into a project plan. Chapter Four gets down to business, covering the nuts and bolts of how to design and develop a mobile-optimized site for your library, reviewing a number of possible approaches and offering tips. Chapter Five covers practical aspects of launching your site and conducting usability testing; it also provides a blueprint for developing a marketing plan to promote your mobile site. Finally, Chapter Six delves into keeping your site fresh and innovative without breaking the bank, the importance of being involved in technology decisions on your campus, and likely sources for keeping up with new developments.

There are many informed voices within the library world—Jason Griffey, Joan Lippincott, and Joe Murphy, to name just a few—contributing to the conversation about the role of mobile services in the context of higher education. In this rapidly evolving landscape, trend-tracking can feel like a full-time job; think of this book as a place to start. In many ways, this is the book we wish we had been able to hand ourselves as we began our exciting and bumpy ride into the mobile web. We hope that it proves useful to you.

mobilelibraries@gmail.com

Chapter 1

It's Never Too Late…But Can It Be Too Early?

Consider Your User Community
Are They Actually Going Mobile?

It's hard to go anywhere these days without seeing someone using a mobile device for all sorts of purposes that don't include phone calls. As Morgan Stanley analyst Mary Meeker reported at the 2009 Web 2.0 Summit, "Mobile Internet usage is and will be bigger than most think" (Parr 2009). But who is using it, and what does that mean for our libraries right now? We assume that the digital natives who comprise our undergraduate population are the most technologically adept of our users. It has been noted that 18- to 34-year-old males are the "most responsive and important demographic group" in terms of mobile services; however, women and older adults represent the greatest potential in mobile adoption (Basso 2009). That means that there is an opportunity to create a space for your library in the daily lives of all types of users you may not be seeing now.

The Educause Center for Applied Research (ECAR) 2009 Research Study on Students and Information Technology found that more than half of respondents owned an Internet-capable handheld device and another 12 percent thought they would acquire one in the next year; almost three-quarters of those who do go online using their device expected their use to increase or greatly increase in the next three years, and almost half of all respondents agreed or strongly agreed that they expected to do many things on a handheld device that they now do on a desktop or laptop computer (Smith, Salaway, and Caruso 2009, 85). This represents a significant shift in terms of how we are accustomed to thinking about access to library services, physically and conceptually. We are already seeing students visiting the reference desk and pointing to a call number or citation on their iPhone or BlackBerry. That gives us some feedback about potential use. What gets utilized and why, particularly when someone isn't actually in the library, will be more difficult to discern.

Ownership of a mobile device doesn't necessarily mean that it is being used to its full capacity. More than a third of undergrads report never going online using their mobile device; on the other hand, almost another third do so daily (Smith, Salaway, and Caruso 2009,

91). For students, cost is a limiting factor, along with speed or performance. For a broader picture, the Pew report "The Mobile Difference," from March 2009, separates ten different categories of information and communication technology (ICT) users into two groups. The first is "motivated by mobility," as demonstrated by their increased online use but also their "positive and improving attitudes about how mobile access makes them more available to others" (Horrigan 2009, 3). The second group feels like they have plenty of connectivity, thank you very much. However, this report was released just a few months before the launch of the iPhone 3GS and the proliferation of smartphones using Google's Android operating system and other platforms such as Palm. Between August 2008 and August 2009, smartphone adoption grew by 63 percent (Reisinger 2009). This competition is translating into more choice in operating systems, features, and carriers. If unlimited data plans continue to be the standard, reluctance is likely to erode further.

Staff Readiness

However, it is not just your students and faculty members to consider in assessing your community. In order to really succeed with outreach to mobile users, your own library staff will also need to be on board—and ideally would be some of the most enthusiastic users. Despite being more likely to rely heavily on technology than the average American, your colleagues may have similar patterns of adoption as your users. The patterns span a continuum from early adopters to those who may be initially reluctant to take on a new technology.

Gamel O. Wiredu provides a thoughtful framework to consider, particularly in the smaller universe of a university campus, noting that "the 'social construction of technology'…suggests that a group's shared judgment of the functional essence of any technology is determined by the context in which it is used and the sense the group makes of it within that context" (2007, 111). Your students and faculty will unconsciously pick up cues on how to respond to new technologies from those around them, and whether it seems to be an accustomed or expected part of their lives. He goes on to argue that a user's attitude towards mobile technology depends on how well it satisfies both an organizational and personal need.

We won't be a deciding factor in whether students and faculty adopt mobile devices, or when, but we might raise awareness by making mobile services available. What we *can* have influence over is how receptive library staff will be. Wiredu is again very helpful here with a discussion on "high flexible computing," which is described as a context where "technology use is enabled by ideal design properties and conditions to support the motive driving the activity" (Wiredu 2007, 122). In other words, the user sees the device as a tool that facilitates a task, as opposed to an object to be wrestled with or not understood.

Why This is Different and Why it Matters

We already know that students are extremely results-focused; if they can get something done right at the point of need, on their phone, that's a big plus. Information is most useful when it's most needed. But mobile devices are blurring boundaries for all segments of our community in terms of place, time, and expectations. Lorcan Dempsey notes that libraries can become mobile-ready in two ways: by optimizing interfaces to work better with mobile devices, but also by restructuring and considering "how to socialize and personalize services" as part of a "shift in emphasis towards workflow integration around the learner or researcher" (2009).

If you assume that everyone is accessing your library and your website in the same way—from a desktop or laptop computer in their dorm or office—you're likely ignoring a big chunk of your community. If you don't offer options, you will miss users who may not know that other service points exist. New technologies can also create a sense of real excitement, something students might not immediately associate with the library. This makes the connection that the library is a place to go for technology and other answers. Because they can be integrated into all segments of our lives, mobile capabilities have the potential to make users feel like they're utilizing a wholly new tool that links them to information and people in a very immediate way. And usability studies and surveys have shown us that feelings are always involved, whether you're talking about libraries or computers.

Tom Haymes has what he calls the "Three-E" strategy for technology adoption: evident, easy to use, and essential. These are all relevant to your user community when it comes to mobile services. To want to use a new technology, it must be evident, and potential adopters need to be made aware of its possible applications; it must be fairly intuitive to pick up; and it must become crucial as a productivity enhancer in daily activities (Haymes 2008). To be the best advocate for providing outreach for mobile users of your library, you'll ideally have reached these thresholds yourself and be able to understand and encourage others who may be uncertain about how new technology will fit into their lives.

Climate and Culture in Your Organization

Offering mobile services will likely mean an accompanying shift in how we conceptualize library access. We can't continue to think of our libraries, particularly academic libraries that serve students and faculty members who have classes and meetings and do research at all hours of day and night, as reflecting even an extended 9 to 5 paradigm. A Pew/Elon survey of leading technologists found that "the divisions between 'personal' time and work time and between physical and virtual reality will be further erased for everyone who's connected, and the results will be mixed in their impact on basic social relations" (Anderson and Rainie 2008). We have been headed in that direction for some time now, but addressing this prospect in a way that incorporates mobile services might mean an accompanying conversation about longer-term plans for your library.

At the very least, paying attention to mobile can be as simple as putting together a stripped-down version of your website, initially. What is most important is awareness—of people and potential. The first step is to start the discussion about how your community is already working, how they are communicating and getting information in other aspects of their lives, and where your library staff sees possibilities to meet them. When BlackBerries and Palm devices were first introduced, they were most prevalent among managers and executives, so chances are your library director or administrators are personally familiar with their selling points. But staff on the frontlines may have valuable observations about how mobile devices have been adopted since then. They offer a range of perspectives on your environment and user behaviors, and incorporating the ideas and knowledge of all staff in brainstorming about what's still a new frontier can only benefit your library's approach.

Conducting a Needs Assessment

In this section, we will present some things to keep in mind when initially considering mobile initiatives. However, the impetus to create new services should not just come from a hunch or a hope; a needs assessment is integral to planning. By undertaking a needs assessment in some form, you will have a better understanding of how to serve your users in this area. You will also have something concrete to refer to in discussions with administrators and staff. Witkin and Altschuld define a needs assessment as "a systematic set of procedures undertaken for the purpose of setting priorities and making decisions" (Witkin 1995). Royse and his coauthors provide a bit more latitude, offering several ways of thinking about it by saying that needs assessment is "a method used to estimate deficiencies…any effort that attempts to determine need…[or] an activity that gages gaps and insufficiencies" (2009).

While assessments can vary widely in method and complexity, there are a few principles that should inform your process, no matter how clear the need for a mobile-services strategy may seem. For more details, the abovementioned books provide good overviews from a social-science perspective. These can be augmented with a library-specific guide (Dudden 2007). There should be three main phases to your mobile needs assessment: planning, collecting data, and analyzing and reporting on findings.

Planning

The first step is to clarify what the goal is and how the results will be used. Keeping in mind that "need" is a relative term and that users and contexts are always changing, you should have a clear sense of the scope and timeframe, who will use this information, and what you have in the way of resources. Do you have two months to develop a strategy, or did you need a mobile site yesterday? Will your report be in the form of an e-mail to team members, a memo to administrators, or a collaboratively written document published on your intranet? Another important element of this phase is to define who your stakeholders are. You'll have primary stakeholders, or the users who will benefit from the services. Are they students, faculty, staff, alumni? Who matters most in terms of mobile devices? You will also have secondary stakeholders, or the staff, administrators, and others who will design and provide your mobile services.

Collecting Data

While "counting is the basic task of needs assessment," there are many ways to use data to assess the viability of a project (Royse 2007, 8). In designing your study, you'll want to decide which tools make sense. Primary data can include user surveys, interviews, or focus groups; you can also incorporate the more unobtrusive means of observation of library users and their devices. Secondary data collection involves making use of already existing information, such as demographic data, analytics information about platforms from your website, or local or national reports about the mobile web for users like yours. No matter your approach, be sure that the questions you're asking will both yield clear responses and cover the gaps in your knowledge of users that you're targeting.

Analyzing and Reporting

The kind of data you collect will dictate in part how extensive your analysis might be. In the most rigorous approaches, you will clean up and code your raw data in order to group it into meaningful categories. In a more qualititative setting, you will be trying to tell the stories that were discovered, ideally grounded in the secondary statistics. When writing up your report, keep in mind who will be reading it and what they want to know about the who and why of mobile services. It is also important to determine ahead of time whether this will be an objective accounting of findings, or whether there will be an editorial presence in the form of advocating for specific mobile services or actions as a result of those findings.

What Could This (Minimally) Look Like?
A Needs Assessment Example
Plan
- Goal: To determine if it is worthwhile to create a mobile site for the library
- Stakeholders: Students, faculty, staff who will use the site; librarians and administrators who plan library services
- Schedule: Decide who will complete which tasks and by what date

Collect Data
- Create a three-question survey (Do you own a handheld device? Have you used it for browsing the internet or text messaging? Have you needed to access the library when you weren't near a computer?), linked on the library website
- Conduct a focus group with library student workers
- Look at Google Analytics information from full-scale library website for past six months for machine/device and browser information

Analyze and Report

- Write a one-page summary of your findings that includes statistics from survey and Analytics along with interesting and representative quotations from the focus group
- Make an "ideal" recommendation for developing a mobile site, as well as a long-term strategy for when to revisit the issue if it is not feasible at present
- Distribute findings to all staff and make full data available

[adapted from Dudden 2007, 78]

Making Mobile Strategic

Once you have some concrete evidence about mobile use in your community, you're ready to move forward. Thinking about where mobile fits in literal terms—what do you have to create, modify, promote, adapt?—also broaches the question of where it fits into your library's strategic plan. Branding and positioning can find a place in even a low-key approach to outreach: what do users expect us to know, and why should they trust us when it comes to information and technology? Being seen as fluent mediators of technology is an intangible that goes a long way to instill trust in your community. They come to the library with questions that straddle the line between content and delivery on a regular basis, and our ability to discern their need is crucial to building word-of-mouth support for the library.

Chapter 2
Fitting the Pieces Together: Integrating with Existing Library Services

We have already covered the importance of exploring how your community might be using mobile devices and undertaking a needs assessment to gather concrete information. This will be especially helpful in explaining to staff or administration why this is a feasible and worthwhile project for your library. The next step is to go back to what should be known terrain for you: what is your library already doing well, and what do users expect in terms of content and services? Mobile isn't just a matter of translating the "big" website to the small screen, but it would be shortsighted not to take into account what is already working. A big difference, however, is that mobile offers the opportunity to make information much more useful by delivering it contextually, which means that your users will need different things from your mobile site. It also presents the chance to reach out to users that you may not be connecting with via the channels your library is already using; more options for access mean a greater likelihood that one of them will make sense for someone who "didn't know we had that" or "never thought the library would provide a service like this."

When it comes down to it, constructing a conceptual framework for your mobile site is a lot like thinking about the landscape of everything your library provides already. You are assembling a mix of externally provided or commercially available content that's integrated with localized information and services, based on your understanding of how your community does its work and what they need. How can you add value through professional judgment? As we always have—by serving as curators, mediators, and guides.

The temptation might be to throw a lot of new and shiny things up there. It catches the eye, after all, and our service-oriented inclination is to give our users everything that could possibly make their lives better. Researchers at the National Library Board of Singapore had to confront this in assessing needs to plan their mobile approach. They had piloted a service that included putting 2D codes—black-and-white squares, sometimes called QR codes, that link users to more information when photographed with a mobile phone—on shelves and posters (Figs. 1 and 2). But it was too much, too soon, and they came to realize that "users didn't want new services but rather wanted to transact traditional library services through a mobile platform" (Pin, Chin and Lian 2009).

Figure 1: QR Code 'in the wild'[1]   Figure 2: Using a phone to take a photo of a QR Code[2]

A similar result was seen in Wiredu's case study of PDA adoption. The custom applications that were loaded onto test devices by the project manager reflected the desired and expected use—and they were a failure. What was successful was the use of the devices for personal tasks at convenient times; in other words, the device allowed users to do things they already knew how to do in a flexible context (Wiredu 2007, 121). Paraphrasing his framework of "an object is something that I have to master, whereas a tool is something I use that makes things easier," you will want to be sure that your mobile efforts aren't forcing users to learn to do new tasks with unfamiliar objects.

How Are Your Users Currently Interacting with Librarians and Library Services?

Before you begin considering how your community is using mobile devices, first think about what's getting used in the library, in person and virtually. You should have some statistics

1. Projeto Sticker Map, *QR code*, Photograph, October 12, 2008, Flickr!, http://www.flickr.com/photos/stickermap/2935588694/.
2. CoCreatr, 1. *find QR code*, Photograph, January 22, 2008, Flickr!, http://www.flickr.com/photos/cocreatr/2211459923/.

to guide you in the form of analytics from your website or reference-transaction totals, as a starting point. For most library users, searching the catalog is paramount, along with accessing account information about charges, holds, and fines. Are there databases with name recognition that are customary first stops for research? Do most of your reference questions come via e-mail, or on IM, or in person? Are librarians with particular responsibilities usually contacted by phone, e-mail, or perhaps through a subject-page widget? Does your library have a news page or blog, or is there a full slate of events and workshops to advertise or coordinate?

These questions will guide what goes into your mobile site and what gets priority on such a small screen. Any evidence you might have of what users are doing on laptops or away from their usual workspaces is also valuable at this stage. For example, once DePaul University's library made an IM widget available on the library website, librarians were flooded with traffic—even from students who were physically in the library. They may not have used that method of contacting the library before, but it fit more easily with their workflow and communication preferences. Knowing this, the library linked to the IM reference widget from the mobile Contact Us page, along with reference and circulation phone numbers, the text-reference (SMS) number already set up to dial, access to an e-mail form, and the staff directory. What else do users want when they're not physically in the library? Hours and locations are an obvious starting place. Survey your community. This can be as simple as talking with student workers or having librarians ask students during instruction sessions about their use of mobile. "Perform contextual inquiries, not focus groups. Go to your users and ask them questions in person, in their context, not yours. They often have a lot to say; listen and keep an open mind" (Fling 2009, 61).

Many reference librarians field inquiries during busy periods about available computers. But showing students a real-time map of how many machines are unoccupied and where provides much better service, and being able to check that information themselves on their phones before even getting to the library takes it a step further. Similarly, while many catalog vendors don't have a mobile-optimized interface yet, our library does offer the option to text a call number from an item record. This has been wildly popular, even without any promotion. This points to exploring more possibilities to incorporate texting information to users, whether it's renewal reminders or short FAQs about resources.

Making the Most of External Services and Resources

These examples bring up the central tenet of allocating real estate on your mobile site: point of need is primary. Whether you can offer something isn't nearly as important as whether, and how, it might get used. Especially in comparison to your full library website, you'll need to be a ruthless editor. And your mobile site presents a similar dilemma in terms of how prominent the concept of the library itself is in the design. As Lorcan Dempsey notes, competition for resources these days points to the need for increased marketing and branding for the library—if no one knows how they're getting to all this amazing stuff, how do we make sure we're still around to provide it?—but some services would ideally be seamlessly integrated so that the library recedes to the point of invisibility for the user (2009).

One of the primary questions is whether to provide a mobile-optimized website that will open in the native browser or develop a stand-alone application, which generally offers more functionality, better speed, and customization for smartphone users. As mentioned in the introduction, the latter is a much bigger project. More institutions are now developing university wide applications (particularly for the iPhone) that incorporate a library presence, though, like the MobilEdu university sites from TerriblyClever for Stanford and Duke. Boopsie is another company working in this area. We'll talk more about finding synergies with other segments of your institutional community in chapter 6. In thinking about resources, be aware that companies often support both a website designed for mobile devices and a downloadable application for the iPhone, Android phone, Blackberry or other smartphone. Your decision to link to mobile sites or highlight apps happens long before the phone is out of someone's pocket: your actual mobile site will generally be comprised of text and links to other mobile-friendly websites. However, this also brings up the issue of the diversity of mobile devices available and the danger of conflating "mobile device" with "smartphone":

> There are still quite a large number of people using small-screened devices for Web browsing. Therefore, if a library's initiative is to provide a mobile version of their site they should develop applications that work on both smart phones and Web-enabled phones equally. However, if the initiative is to build innovative applications that use functionalities that are included on a family of devices, then developing an application that exploits the de-

vice's capabilities to enhance a user's experience is a better solution. (Griggs, Bridges, and Rempel 2009)

If you want to recommend relevant applications for your community, consider a page on your full site that explains a bit about the mobile environment and points to useful academic apps, which can then be downloaded for the particular device; several examples of this approach will be seen in chapter 4. You can use the guidelines in this chapter, along with your own testing, to winnow potential recommendations. For example, many academic-library websites already include their institution's course-management login as a convenience; be sure to try common tasks on yours before linking to it. Don't undermine the trust of your users by pointing to (that is, implicitly recommending) a service that doesn't perform the way it should and isn't ready for prime time yet. On the other hand, do you introduce students and faculty to related resources like Zotero for citation management in instruction sessions? Do you think they'd also like to know about apps like Margins, which saves annotations you make linked to citation information, or Papers, which allows you to search, read, and sync scientific and technical research literature?

A few big names supporting mobile sites and applications
Amazon (http://www.amazon.com/)

Fig 3: Amazon iPhone app   Fig 4: Amazon app menu   Fig 5: Amazon mobile site   Fig 6: mobile site menu

Try to access Amazon, seller of everything from books to music to toothbrushes, on your mobile device and you're automatically redirected to a stripped-down version: search for an item, get some recommendations, or browse categories by top sellers. Oh, and did you want a Kindle? (Figs 3-6)

Flickr (http://m.flickr.com/)

Fig 7: Flickr iPhone app   Fig 8: Flickr iPhone app   Fig 9: Flickr mobile site   Fig 10: mobile site menu

Flickr, a photo-sharing and management service, lets you access your own or friends' photos as well as browse through millions of uploaded photos. Although image-heavy mobile sites are unusual, this is their stock in trade. The application allows the user, once authenticated, to access personal information (photostream, activity, uploads, contacts) much more simply, and integrates the native Camera app for easy uploading (Figs 7-10)

NYT (http://mobile.nytimes.com/)

Nobody wants all the news that's fit to print on a mobile screen, so the NYT gives you the basics, relying on categories to get you where you need to go. The app allows you to quickly review the latest stories, the most e-mailed stories, and any stories you might have saved. (Figs 11-13)

Fig 11: NYT iPhone app   Fig 12: NYT iPhone app   Fig 13: NYT mobile site

Don't Reinvent the Wheel

As more and more vendors realize the opportunities that mobile presents, there will be more options to include on your site. This creates an ongoing process of trying to discern what's available and what's necessary—developing a mobile website does not involve one-time decisions. Be aware, though, that your users might very well find their own way to resources, and likely ones that don't require authentication like those that are behind the library's pay wall; this is more of a navigation hurdle on a small screen, although the University of Michigan has released a bookmarklet to their users that will add the proxy information to any page with one click. Will your mobile users look for anything more than Wikipedia for quick, good-enough answers? If so, or if you have strong philosophical reservations about Wikipedia, point them to Britannica for ready reference. The usa.gov site is often a happy surprise for users at the reference desk, so consider including their mobile site. WorldCat and PubMed are heavily used and well-recognized databases at our library, and their mobile-optimized sites made them strong candidates for inclusion. Do you have expensive databases that you'd like to promote more, or do you subscribe to multiple databases on the same platform? Those are all possible candidates.

There are other ways to leverage already existing services. While we won't cover too much in the way of SMS services, they can be an important part of your mobile strategy. The Hong Kong Institute of Education Library is repurposing the reminder function in Google

Calendar to send text alerts about overdue, pick-up, and recall notices (Mak 2009). The Skokie Public Library (IL) is doing the same thing but with enhanced functionality using a suite of services from Shoutbomb (Greenwalt 2009). You can also find ways to support or piggyback onto institutionwide initiatives. Southern Methodist University (TX) gave their incoming first-year classes iPhones for the past two years. Houston Community College (TX) experimented with two sections of the same course; in one, students were given iPhones, while other students accessed course materials online in more traditional ways. The mobile-enabled students utilized the class resources more and said they were more engaged (Young 2009). Linking to course pages from the library mobile site not only raises awareness of what's possible but also keeps the library identified with resources, no matter where they live or are accessed.

Worksheet: Getting Your Toolbox Ready

- List what is getting used most in your library already
- Next, list what seem like the highest priorities for mobile
- Flag the overlap in these two lists (similar to the intersecting part of a Venn diagram) as your top resources
- Finally, put together your "blue sky" list of things you'd love to be able to do

Chapter 3
What's the Plan? Developing a Project Plan and Proposal

Now that you've taken some time to consider whether developing a mobile version of your website makes sense for your community and for your library, and identified how best to hook into existing services, it's time to put together a plan of action. That's right: project management (PM). However businesslike and off-putting (or boring) it may sound, PM is a crucial skill: if you manage the project, it won't manage you.

A recent study discovered that while courses on library management are offered by nearly all accredited LIS programs, only two offered courses specifically on PM (Winston and Hoffman 2005). Luckily, for those of us who weren't fortunate enough to study this as part of our LIS program, there are many excellent sources available, as well as formal certification or degree programs; what we'll present here is just a quick overview to get you started, focusing on PM in libraries.

To begin at the beginning, a project is "a planned undertaking and set of related activities that have a defined beginning and end" (Giesecke and McNeil 2005, 147). Reflect on the many "projects" you may have encountered in the workplace, existing in a sort of half-life, handed down from librarian to librarian over a long period of time yet never finished—not a fate any of us would like for our mobile sites, to be sure! These poor unfortunate zombie-projects probably didn't have the benefit of a carefully considered plan, which can "save time and money by streamlining the work, identifying problems beforehand or by discovering the project is not worth the effort… at the present time. A project may sound great when it seems to be cheap and easy, but…[not when] revealed to be expensive and difficult" (Barba and Perrin 2009).

Projects may be change-driven, aiming to increase effectiveness or efficiency; crisis-driven, responding to a specific problem that needs to be solved or corrected; or market-driven, responding to evolving needs of the user population. Depending on the outcome of your environmental scan, you may consider your mobile website project to be in any one of those categories. Regardless of which, you can think of the project as having four major steps, or phases (Wamsley 2009):

1. Initiation
2. Planning

3. Execution
4. Shutdown / Evaluation

Phase One: Initiation

The first step will be to identify a project manager—we'll assume that's you. Since you will take responsibility for supervising the project and shepherding it to its completion, it is important that you also have the authority to make necessary decisions regarding staffing, scheduling, budgets, and project scope. In larger organizations, it may be helpful to identify an administrator as the project sponsor, who can serve as a resource to and champion for the project manager and team.

Next, you need to make some decisions about scope. Yes, you're building a mobile site, but the extent of services you offer can vary considerably. What exactly does your plan encompass? (And, even more importantly, what will you not be doing? Resist the urge to stuff ten pounds into a five-pound bag!) At the same time, take the opportunity to simply and clearly articulate why you consider developing a mobile website to be a worthy investment of time and resources for your library, addressing what you might call the "so what?" factor: Why should we do this? What is the ultimate value to our staff and our community? In his pragmatic and helpful book *Observing the User Experience*, Mike Kuniavsky discusses issues pertinent to project planning (2003, 507). He proposes the following as the key questions:

- Why is it being done?
- What does it do?
- Who cares about it?

These answers will be useful later as you frame your goals and objectives. There is one last question to answer before presenting your proposal to administration: how long do you estimate this will take? In step two, we'll return to scheduling in more detail, but for now, answering what, why, who, and how long it will take (at least a ballpark figure) should be enough to seek approval to move forward to the planning phase.

You'll also need a team, which takes some careful consideration. What is the ultimate goal of your project? If you see this as a pilot project or a proof of concept, you'll want to keep the team quite small, both to minimize impact on staff time and to remain quick and nimble. Limiting the scope of your project may actually result in more opportunity, especially in risk-averse environments; given this, author Aimee Fifarek cautions against going overboard (2007). Kuniavsky also points out, regarding small projects, that "when they fail, small projects are not seen as severe setbacks. When they succeed, they can be used as leverage for larger projects" (2003, 509). The software company 37signals has based their company and business philosophy around the idea of focused simplicity; in their book *Getting Real* (focused and simple, unsurprisingly) they recommend a team of three (2006, 35). If you are viewing this first effort as a more comprehensive, long-term undertaking, you need to take special care that important stakeholder groups are represented. In many cases, this will dovetail neatly with identifying staff who can contribute the necessary skill sets, and who are enthusiastic about the project. In either case, recognizing that a high-functioning, productive team must be actively cultivated and supported by the project manager will contribute to your success.

In some cases, you may have the autonomy and infrastructure to simply embark on your mobile site without explicitly seeking administrative approval. This is also more likely to be the case when envisioning your efforts as a proof of concept, since such a project will by definition be extremely limited in scope. While it can be exciting to strike out boldly into a new endeavor, be sure that doing so doesn't endanger the future of your project. Furthermore, don't underestimate the value of written documents in formalizing and building consensus, in sharpening your own vision of what is to be accomplished, and in beginning the very important process of project documentation. You need to take into account the various implications of launching a new service, and formal written documentation goes a long way in that regard. Frank Cervone points out that "fostering an environment of successful project leadership comes as a result of earning the respect of colleagues throughout the organization" (2008, 200). A written project charter need not be overlong but will clarify boundaries and expectations (Wamsley 2009). Of course, as with anything, play it by ear within your organizational environment. Overformality has killed many a project, but then, so has overfamiliarity. A simple page-long document may be sufficient to get things started.

Phase Two: Planning

Once you have gotten the go-ahead from your administration and selected your team, it's time to put together a more detailed plan. A number of components of the overall management of the project need to be addressed in planning: scope, time, cost, quality, human resource, communications, risk, procurement, and integration (Cervone 2007). While all are important to consider, for this project perhaps the most crucial are scope, time, communications, and integration.

Scope has already been addressed in the project-initiation phase; at this point, the answers to the twin questions "What are we setting out to do?" and "Why are we doing it?" can be transformed into goals and objectives. You may already be familiar with the concept of SMART objectives from strategic-planning initiatives or your own reading. SMART stands for Specific, Measurable, Agreed-upon, Realistic, and Time-bound (Wamsley 2009). Fuzzy objectives can neither be accomplished nor measured.

Time waits for no plan. Good scheduling includes more than just a timeline of deadlines for the work itself. Take into account the other regular responsibilities of your team; ebbs and flows in general "busyness" of your library; coordinating hand-offs between staff members or departments (for example, do you have access to servers, or will you need to arrange for

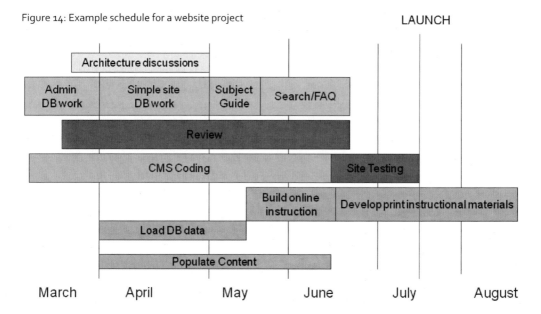

Figure 14: Example schedule for a website project

another person or department to complete certain parts of the work for you?); all necessary resources in addition to staff (access to systems and technology, financial requirements); and dependencies and concurrencies within the project itself. How best can the work be sequenced? Especially as the number of people or units involved increases, it can be helpful to develop a matrix of task assignments and deadlines (Wamsley 2009; Marill and Lesher 2007). (Fig. 14)

Once you've come up with a schedule, be sure to pad it a bit to correct for the inevitable, but unpredictable. If everything goes perfectly, you'll finish early, but at least you won't go over schedule!

Communications encompasses interchanges within the team, as well as reports to administration and to stakeholders within and outside the library. How often will you need to touch base with team members to be sure that questions or issues are addressed as they arise? Too many meetings will impede actual work; too few may result in delays due to procrastination or effort wasted due to misunderstandings. What level of updating is expected by your superiors? How will you communicate progress to the larger groups of stakeholders? What venues will be most effective for each? In today's technological environment, the range of options for collaborative web-based applications or workspaces can be almost overwhelming. This book, for example, was completed as a small-team effort using a combination of periodic in-person meetings, e-mail, instant messaging, and shared drafts using Google Docs. Administration may prefer regularly scheduled meetings, or short written reports. For communication to wider groups, it will probably be most appropriate to formally present progress and solicit feedback at meetings scheduled to coincide with project milestones such as settling on a design, enabling access to a test site, and immediately prior to launch. Another important aspect of communication management is making sure that the existence of the project is generally known, so that anyone with questions may direct them to you, the project manager. This will do a lot to circumvent unprofitable conjecture, rumor, or anxiety about your project.

Integration really just refers to change management. Even with a small, tightly scoped project, you can't predict everything. Challenges and opportunities will arise throughout the execution phase; sometimes, even totally new technologies will suddenly emerge. In an environment as volatile as the mobile web, you can bank on this happening prior to your launch. How much latitude will you have to adjust for these changes during implementation? How will you decide which will be beneficial to your initial mobile-optimized site, and which must be deferred to the next update or version? From where or whom do you need to seek approval to make modifications as you go along?

This is also the time to consider two other important ideas: risk management and assessment. They are not fun, they're not sexy, they're not shiny, but your mobile site, iPhone application, or any other future Next Big Thing will probably not stand much of a chance of being truly successful if you fail to grapple with them in the planning phase.

Perhaps a less intimidating way to think of **risk management** is to think through a few scenarios. What potential pitfalls can you identify prior to beginning? What is your Plan B for dealing with these issues if they should arise? What is the chain of command for escalating problems, particularly those that were unexpected?

Finally, as budgets tighten, **assessment** is more and more a focus of attention in libraries. It doesn't just have to be a way to keep the wolf from the door, though. For an excellent brief introduction to the topic, with helpful references, try Lisa R. Horowitz's recent article, "Assessing Library Services: A Practical Guide for the Nonexpert" (2009). Not only is it important to be able to justify your decisions through metrics that show that time and resources were well invested, and to make adjustments in areas where data indicate improvements are possible, but the construction of a thoughtful assessment plan allows you to more easily measure your success. There is also the pragmatic issue of the care and feeding of your project later—who will mind it once it is up and running? What impact will there be on existing workflows? The more of these questions you can address, the better you have positioned yourself for the future.

It's all very well to talk about metrics, but the question remains: which metrics do you need to collect? And let's be honest: what does "metrics" mean, anyway? Clearly, some measures tell us how much or how many. How many users, how many clicks, how often a particular page is visited, what browser and operating system was visited, how much time do users spend on your site? Kuniavsky says further that "metrics abstract aspects of the user experience in a way that can be measured and compared…[they] begin with goals" (2003, 517). Time to refer, once again, to your stated goals. While the quick list of aforementioned metrics is probably a good place to begin, you may find that there is additional, more specific data you'd like to monitor, collect, and analyze.

Having decided what you plan to collect, how will you collect it? While system administrators at some libraries and institutions may maintain server logs of data, Google Analytics is an excellent tool all librarians should consider for its powerful reporting and its ease of setup and use (Andrew 2009). (Fig. 15) Even better, they recently increased reporting for mobile applications (Papp 2009).

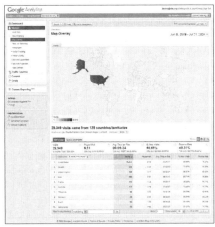

Figure 15: Example of the Google Analytics dashboard and map overlay, via ALA Connect Flickr[3]

Another facet of assessment specific to website projects relates to user testing, which will be discussed in more detail in chapter 5. In the context of planning, you will want to decide which methods to employ (usability testing, focus groups, surveys, heuristic evaluations, etc.), when to schedule them, who will take responsibility for conducting the work, and how to report the results.

Phase Three: Execution

…And go! Now it's time to actually start work on the project. All the effort you invest in planning should allow you to focus your efforts on your responsibilities as a project manager, including making sure everyone stays on task and on schedule, maintaining communication

---

3. ALAStaff, July 22, 2009, ALA Connect Statistics (2009-07) [set], Flickr! http://www.flickr.com/photos/alastaff/sets/72157622789077553/

as laid out in your plan, and addressing issues as they arise. As adjustments or major altera-tions occur, make sure that project documentation is updated. It's easier to make corrections as you go along than to try and retrospectively log changes later.

An important caveat, especially in a chapter devoted to the virtues of planning: however bul-letproof the plan appears to be, it is rare that any project goes exactly to plan; when it doesn't, be kind to yourself and your team. Expect to iterate—sometimes after completion, and sometimes as a part of the process of completion. Particularly if this is the first time you are embarking on a project of this kind, or your first experience as a project manager, you can almost definitely expect to make mistakes or for the unexpected to arise. Most will likely be benign, a useful part of the learning process; there may also be some howlers, but those at least furnish useful lessons on what (or what not) to do (n.b.: the authors have discovered they are also sometimes an excellent source of material for self-deprecatingly humorous stories, after the fact).

Phase Four: Shutdown / Evaluation

Congratulations, you completed your project—your mobile website is up and running. Be-fore you archive everything and move on to your next challenge, there are a few important activities to undertake. Evaluate both the end result and the process itself. For each, answer these questions: What went well? What could have gone better? Evaluation also includes completing the assessment plan you put in place earlier, including collecting and analyzing the statistics, and revisiting your ongoing plan for data collection and analysis. Will you need to make any adjustments given your evaluation?

Document the process and create supporting materials for ongoing maintenance. In some cases, you may retain the responsibility for overseeing the mobile site for the long term. In any case, for yourself, for others at your institution, and for those who come after you, documentation will save a lot of time spent wondering about the hows and whys of decisions and processes. As the site grows in size and complexity, you may need to involve more staff. Preparing a "road map" of your project as promptly as possible after its completion will ease transitions and ensure that important information is recorded (Barba 2009).

Report to your constituencies: the project team, stakeholders, administration, and end users. Clearly, these are very different groups with very different needs. While the bulk of the information you present to each may be largely the same, the format and tone will not be.

Acknowledge the work of the project team and the contributions of others who assisted or supported the project by creating a space in which to celebrate your success. This can be something as simple as a public acknowledgment via e-mail or in a meeting, or something more elaborate (and possibly involving edibles). This provides closure not just for the project team but also for the staff.

Worksheet: Putting Together a Project Proposal Document

University Library Mobile Website Project Proposal

Summary
In three sentences or less, what are you proposing to do? [Tip: if you write this last, it'll be easier!]

Statement of Purpose
What is the project aiming to achieve?
Why is it important to achieve the stated aims?

Oversight
Who will be involved in working on the project and what are their roles and responsibilities? How do these responsibilities interact with existing job duties?

Project Timeline
When will the specific elements or aspects of the project proposed be put into effect? Remember, this is your ballpark figure. It may be helpful to think broadly, in the context of academic calendars.

Drafted by [staff]: [date]
Submitted for approval: [date]
Approved by Dean: [date]

Chapter 4
Let's Do It: Building the Site

Now you have a plan for the project overall, and it's time to build the site. Where to begin? Cameron Moll neatly lays out four strategies for approaching your mobile site in his book *Mobile Web Design* (2007):

1. Do Nothing
2. Reduce Images and Styling
3. Use Handheld Style Sheets
4. Create Mobile-Optimized Content

Clearly, if you're reading this book, you're not leaning towards option one, but interestingly, his recommendation for the most successful mobile site is to choose between "Do Nothing" or "Create Mobile-Optimized Content." We agree with him. Later in this chapter we'll touch on options two and three, so you can decide for yourself, but for now, we'll start with some issues related to option four. After all, if you're going to create mobile-optimized content, you're going to have to do some designing.

Designing the Site

There are many excellent books, print and online periodicals, and blog posts written on web design by experts in the field. And since this is intended to be a pragmatic hands-on guide, written by nonexperts for nonexperts, we will focus on a few key considerations for good design in any website, whether mobile-optimized or not.

Design for Your User

This may seem obvious but apparently isn't, based on the seemingly endless number of publications imploring companies and institutions to keep the user in mind as they design. Call it user-centered design, customer-centered design, experience design, participatory design, "getting real" or just plain common sense: who is your user and what is she trying to do at your website? How can you make that as easy as possible? Everyone who comes to your library's website is there in the pursuit of another goal, and the library website is just one stop along the way. The easier your site is to use, the more transparent it becomes, allowing the user to concentrate his or her energies on the heavy mental lifting of the research project, literature review, or information need that spurred the visit. In the case of mobile devices, the constraints of use (smaller screen, limited inputs, wildly variable environmental conditions) only exacerbate the situation. Simple solutions require complexity of thought on the part of the designer, which reduces the amount of cognitive work shifted to the user.

So what does that mean you need to do? First, refer to your needs assessment. What does it tell you about your user population? What are the main information needs of your audience? How did you wrap that into the scope statements during the planning process? Then look at the list of services and functions you drew up in chapter 2: that is the nuts-and-bolts content of your mobile site and can quickly and easily become the beginnings of your working site map. Don't get carried away, though; keep it simple. You can always add more features or functions later. Some good rules of thumb can be found in *Smashing Magazine*'s top five mobile web design trends for 2009 (Snell 2009):

- Simple options (in tandem with the last trend listed here, offer only what is necessary)
- White space (easier to read, easier to manipulate)
- Lack of images (quicker download times, less clutter)
- Subdomains rather than separate domains (m.library.univ.edu, library.univ.edu/m/, or library.univ.edu/mobile rather than library.univ.mobi)
- Prioritized content (as we discussed in chapter 2, curate your content and provide access to only what's key)

This last point in particular jibes with something Steve Krug so rightly points out: people don't read, they scan, and this is even more pronounced on a mobile site, viewed on a very small screen (2006, 21–25). Dakota Reese Brown goes further, saying that "if browsing the Internet from a desktop is regarded as a scanning activity, then browsing the Internet through the adaptive lens of a mobile browser might best be described as a squinting activity" (2009). Add to that awkwardness of input—the jabbing "fat fingers" problem of the touch-screen mobile device, or the endless tabbing of earlier models of mobile-enabled phones—and you'll want to boil your content down to the absolute minimum (Nielsen 2009).

Given those considerations, what should it look like? The answer: it depends. Let's start by looking at a few library sites and some widely used commercial sites to get a sense of the breadth and range of solutions. (Remember, you can access all the links we reference throughout the book, as well as many more, via our Delicious account at http://delicious.com/mobilelibraries)

Libraries

These are just a few examples of mobile-optimized library sites. (Screenshots are exclusively from the iPhone in this section; there just wasn't that much difference for these sites between the smartphone displays. Also, if you know of other notable examples, please send us an e-mail—see our e-mail address, available in the introduction.)

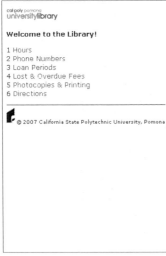

Figure 16: As viewed on iPhone

California State Polytechnic University, Pomona (http://www.csupomona.edu/~library/mobile/)
If you start doing much reading on mobile site design, you'll doubtless come across an article where the author mentions that a site "degrades gracefully," meaning that it displays well both on a smartphone and on less sophisticated web-enabled devices. This is a good illustration of how to keep a site simple, basic, and clean.

College of Saint Benedict— St John's University (http://www.csbsju.edu/library/mobile/)

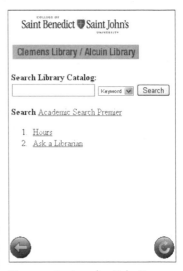

Figure 17: As viewed on Palm Pre

Most important in the design of any site is making the choices that work best for your community. This site is similar to the previous one in its simplicity but places more emphasis on research resources.

Figure 18: Full site, as viewed on iPhone

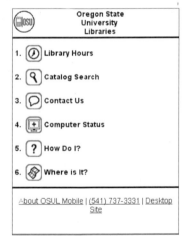

Figure 20: As viewed on iPhone

Indiana University Bloomington (http://libraries.iub.edu/m/) IUB is notable for its particularly nice handling of allowing users of smartphones to access the mobile site via a notation in the banner graphic of their standard site at http://libraries.iub.edu/

North Carolina State University Libraries (http://www.lib.ncsu.edu/m/) It's also useful to look at a really well-designed site optimized specifically for smartphones. NCSU has done a great job of creating great user experience on an iPhone through use of icons and styling. More information about their project: http://www.lib.ncsu.edu/m/about.html

Oregon State University Libraries (http://m.library.oregonstate.edu) A simple design emphasizing a strong use of color and iconography. More information about their project: http://osulibrary.oregonstate.edu/about_mobile

Figure 19: As viewed on iPhone

University of Virginia (http://m.lib.virginia.edu/) UVa, like NCSU, takes advantage of styling capabilities for smartphones. They also provide a search box labelled 'Search the library', which searches their catalog; this brings up the importance of making labels both brief and clear. More

Figure 21: As viewed on iPhone

Figure 23: As viewed on iPhone

information about their project: http://www2.lib.virginia.edu/mobile/

Yale Cushing/Whitney Memorial Medical Library
(http://www.med.yale.edu/library/m/)
A third example illustrating how asking 'What choices are the best for your community?' will determine what content you prioritize.

Figure 22: As viewed on iPhone

Orange County (FL) Public Library
(http://m.ocls.info/)
OCPL emphasizes news but allows easy access to other content using top navigation. Notice they highlight Videos—they have a robust presence on YouTube.

Skokie (IL) Public Library
(http://www.skokielibrary.info/mobile/)
While we've emphasized the importance of editing your content to reflect your highest priorities, that doesn't mean you can't have more than five links. Skokie PL gives their patrons a slightly longer list of choices than some we've seen, but it is by no means overwhelming.

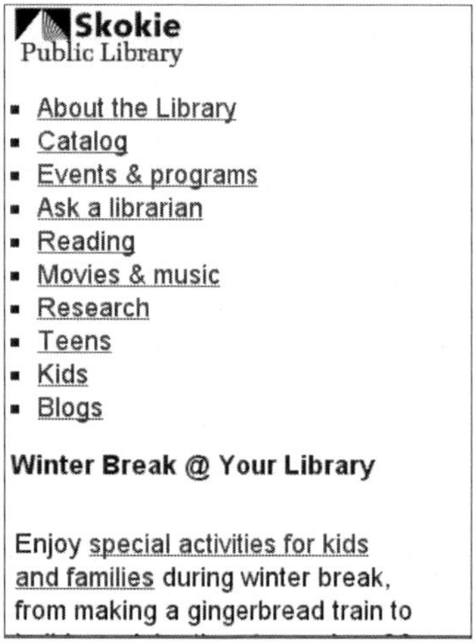

Skokie
Public Library
About the Library
Catalog
Events & programs
Ask a librarian
Reading
Movies & music
Research
Teens
Kids
Blogs
Winter Break @ Your Library
Enjoy special activities for kids and families during winter break, from making a gingerbread train to

Figure 24: As viewed on iPhone

It's useful to review examples of library mobile-optimized sites, but don't feel you need to stop there. Commercial sites can offer plenty in the way of inspiration. How are they providing solutions to their customer base?

Airlines

There's a lot of stress inherent in travel—Is my flight on time? Can I check in from here? Do I have to print a boarding pass? Can I reschedule my flight?—and being able to answer these questions on the go is a boon. Students certainly experience some of these same levels of stress related to completing their assignments, meeting deadlines, and conducting research using library websites, whether in a mobile or desktop context.

Southwest (http://mobile.southwest.com/)

Welcome to Southwest
SOUTHWEST.COM
Book and Modify Air
Check In
Flight Status
Flight Schedules
Rapid Rewards

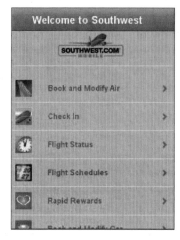

Figure 25: As viewed on iPhone

SOUTHWEST.COM
MOBILE
Book and Modify Air
Check In
Flight Status
Flight Schedules
Rapid Rewards
Book and Modify Car
Contact Southwest
View Full Site
Help
More

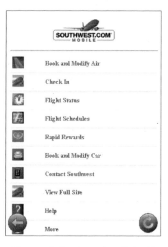

Figure 26: As viewed on Palm Pre

American Airlines (http://mobile.aa.com/)

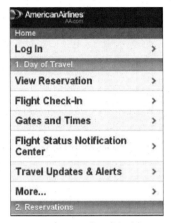

Figure 27: As viewed on iPhone

Figure 28: As viewed on Palm Pre

Encyclopedias

When people go to an online encyclopedia, they want to look something up or might just want to browse information.

Encyclopedia Britannica (http://i.eb.com)

Figure 29: As viewed on iPhone

Figure 30: As viewed on Palm Pre

Figure 31: Search dialog

Note: There is no visible search box, just a button. Bear in mind there's no additional click cost to bring up the keyboard, where the user has to touch a text box or the search button.

Wikipedia (http://en.m.wikipedia.org/)

Figure 32: As viewed on iPhone   Figure 33: As viewed on Palm Pre

Their mobile site has a prominent search box, and a randomly selected featured article. Wikipedia also has a (free) downloadable iPhone application, which sticks to the same layout with the added function of search history.

Standards and Accessibility
"Standards" and "accessibility" are terms that are frequently bandied about, and much less frequently defined. As librarians, we already have a firm grasp on the importance of controlled vocabulary, as well as a healthy sense of how inflexible it can be. Web standards are more or less the same thing. The World Wide Web Consortium (W3C) maintains a set of conventions for HTML, CSS, AJAX, and other building blocks of websites with an eye to enabling simple marked-up pages that render consistently across browsers and devices. In addition to being easier for browsers to interpret, standards-compliant sites are also easier (and cheaper) to maintain. The good news about standards and accessibility is that they

go hand in hand (Web Accessibility Initiative 2008). If you use standards, you will almost certainly have a site that will also satisfy the four principles of accessibility, as defined by the Web Accessibility Initiative:

- Perceivable ("can't be invisible to all of their senses");
- Operable (can be manipulated);
- Understandable;
- and Robust (adapts to evolution of technology) (Web Accessibility Initiative 2009).

Happily for us, an accessible site, in addition to being the right thing to do, is very likely going to be much more usable and friendly on a mobile device.

Picking and Choosing

To undertake the design of a full-blown mobile website, there are a number of steps. Some of these we've undertaken already, albeit in a limited way: considering user needs; reviewing services and offerings to decide what should be included; defining the site's objectives and scope; setting up a team and putting together a project plan, schedule, and budget. An important part of any web-development project is pulling together your functional requirements, which can be thought of as falling into two major categories: what does the site need to include (content), and what does it need to be able to do (coding)? Requirements are in many ways just a technology-specific outcropping of your project plan and scope statements.

When you're building something yourself, or when you're building something that has a limited scope, laying out your requirements might seem less important than it might when tackling a site encompassing many services, departments, people, and pages, and involving a number of people. Still, as when doing research, lack of clarity can lead to a lot of extra work. If you're not sure of what you're looking for, how will you know when you've found it? In this case, if you're not sure what end result you desire, how will you know when you're done?

The importance of being clear on what you plan to do and how you envision it working cannot be overstated when working with developers. Functional requirements serve as the "contract" for your project. No one else lives in your head, and if your site is not clearly defined from the start, you might end up with something quite different than you imagine.

Jesse James Garrett says:

> In the absence of documented requirements, your project will probably turn
> out like a schoolyard game of "Telephone"—each person on the team gets
> an impression of the product via word of mouth, and everyone's description
> ends up slightly different. …By establishing concrete sets of development
> requirements and stockpiling any requests that don't fit those requirements
> as possibilities for future releases, you can manage the entire process in a
> more deliberate and conscious way. (2002, 63, 65)

Take your list from the worksheet at the end of chapter 2, which is, more or less, your requirements list. We're going to use it to put together our site. If you've ever done storyboarding, where you visually lay out a sequence, our next step will be a bit like that. We need to take our list of ideas and make them into pages. There are a lot of ways you can do this, but we like Post-It™ notes. They make it easy to move things around, they provide a tactile, concrete element to the whole process, and their essential impermanence tends to take the edge off and give people the freedom to try a lot of ideas without getting stuck on the first suggestion.

On each Post-It™ note, write the name of the service or page to which you'd be linking. You might consider using different colored Post-It™ notes for the different categories of items (essential, would be nice, external, blue sky). If the item requires anything "special" or has any dynamic behaviors, note that; it will generally imply that you have some coding to address. Now you just need to place the notes in a sequence or order. What comes first? Are you organizing by priority or in alphabetical order? As you consider each item, ask yourself if it's doable now, or complex enough that it needs to wait for version 2.0 and be added later? It's been said before, but it bears repeating: "Since mobile devices are likely to be the smallest screens in a user's experience, the design of mobile experiences must accommodate the user's varying commitment and distributed attention" (Reese Brown 2009).

Designers often use a technique called wireframing for early prototyping. With a mobile site, you're probably not going to need to go much beyond this level of design. Wireframing is a lot like storyboarding; you just want to begin with a rough sketch of the layout. It becomes a "wireframe" when you mock it up in a graphics application, or code it into very simple

HTML. Take the list you've created and sketch it out in a box with the relative dimensions of the screen size you're designing for. If you have a larger team, a bigger budget, or a higher skill level you can certainly plan to launch a site optimized for different devices through use of stylesheets or redirect scripts or such, but you'll probably want to begin with a single, simple site that will render consistently for almost everyone and go from there.

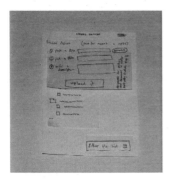

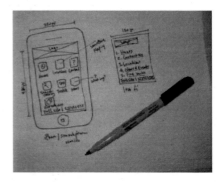

Figure 34: Designing with Post-Its[4]   Figure 35: Wireframing[5]   Figure 36: Wireframing—sketches for an iPhone app

Remember that you need to accommodate two different interaction styles, touch (smartphones) and scroll (all the others). Also, the screen sizes vary considerably. While the iPhone and Palm Pre have an overall screen size of 320x480, and the various BlackBerry model widths range from 240 to 480, many web-enabled phones have a maximum screen width of 120 pixels. Keep this in mind when you consider the width of images and even of your text. At this stage, you'll also plan for important things like useful, consistent footers. Along these lines, Oregon State University drew up a tremendously useful list of 10 mobile design recommendations, which are detailed in "library/mobile: Tips on Designing and Developing Mobile Web Sites" (http://journal.code4lib.org/articles/2055)—freely available, and a

---

4. Yandle, *Media Chooser*, January 21, 2008, Flickr!, http://www.flickr.com/photos/yandle/2208732061/.

5. Rob Enslin, *SketchBook Mobile Express*, September 21, 2009, Flickr!, http://www.flickr.com/photos/doos/3942284566/.

must-read, by the way, for the entire process (Griggs, Bridges, and Rempel 2009). There are other particularities and details you'll need to negotiate, but we'll cover those a bit later in the chapter. Right now you want to get a basic layout that you can work from.

Developing the Site

Once you have a sense of what the site will include and how you will lay it out, it's time to move on to the actual development, or coding. How you decide to approach this will depend in large part on your organization, your budget, and your skills. Let's work through this as a decision tree.

1. Do you have an in-house IT department that maintains your web presence? Alternatively, is your web presence entirely overseen by campus IT?
    NO? Skip to 2.
—>**YES.** In this case, you need to talk to these people. In fact, you have probably been talking to them since at least chapter 2. If you haven't, we'd advise you schedule an appointment immediately because it will be impossible to construct a relevant, usable project plan without the entire project group!

2. What's your budget for this project? Do you, in fact, have a budget? Could you potentially be awarded grant funding through your campus, institution, a consortia, or other group?
    NO? Skip to 3.
—>**YES.** If you don't have the staff or the skills personally, but you do have flexibility in your budget lines, obviously the easiest thing will be to just pay someone to do it. Grants can be very helpful to get funding for one-time projects—like the development of a site—that can then be maintained with existing staff. Might outsourcing be the right choice for you? There are external companies, like Boopsie, to whom you can contract the work. You might be able to arrange this internally as well, through hiring a part-time worker or an independent contractor, or funding a graduate assistantship. In addition to the actual cost of the salary, there are the additional invisible costs in terms of your time in hiring, training, and supervision. Be sure to consider those as well.

3. Do you use a content management system? Is your site styled with CSS (Cascading Style Sheets)?

   NO? Skip to 4.

—>**YES.** If your library's website is built within a content management system (CMS), you will almost certainly already be using CSS to style it; in fact, the likelihood that you are employing CSS in your site, however it was built or is maintained, is extremely high. If you want to think of HTML or XHTML as creating the basic structure of a web page and providing its content, then cascading style sheets (CSS) determine its look and feel: fonts, colors, spacing, even layout. Depending on the product you use, you may or may not have an option to enable an existing "mobile" stylesheet, which is sort of like applying a template. Your CMS may instead allow you to install a plug-in, or you may be able to create a mobile stylesheet without much ado (see question 4 for more on that). This might make it a very simple matter to create a few pages within your CMS to which you can apply the mobile template, and voila! you have a site running. If you have a commercial CMS, check the documentation, or contact the vendor. For open-source CMS, the web is a rich source of help.

Alternatively, depending on the structure of your current site, you could write a separate stylesheet for mobile devices. That is, it is technically possible to apply such a stylesheet to the entire site. As a general rule, library websites are fairly complex; it might be difficult, time-consuming, or even impossible to try and write a stylesheet that would "optimize" your site for a mobile device. The real question here is whether your users really need all the information on your library site in the context of mobile usage. Even if you could give it to them, would you be doing them a service? After all, "do no harm" is a good rule of thumb for every profession. Consider this: "First and foremost, simply transferring a full-sized computer application to the mobile environment almost always results in a suboptimal mobile experience... Mobilizing an application means reconsidering the entire purpose of the application, not just changing display technologies or interaction nuances" (Ballard 2007, 70-71)

4. Can you write HTML/XHTML? Have you ever written a stylesheet?

   NO? It's worth a try anyway! You can also skip to 5, but we'd recommend you focus your time here.

—>**YES.** Ah, the Old-Fashioned Way: coding by hand. To some this may sound overwhelming, but remember the advice given in regard to mobile sites is to keep them simple, with minimal content and minimal styling. You want a site that opens quickly, that isn't cluttered, and that addresses the primary needs of the mobile user. Just because it is possible to include lots of fancy bells and whistles doesn't mean that you need to feel obligated to do so, especially since you want to test how users respond to the design on your first time out.

There are a couple of things you'll need to know that are particular to mobile-enabled sites. It should be noted that there is a bit of a debate on the preferred markup language version: the World Wide Web Consortium (W3C) says XHTML Basic 1.1, others point to XHTML Mobile (sometimes noted XHTML-MP) 1.1 (Passani 2009). How do you decide? If you want to be able to support a wider array of "lower-end" (aka not-smartphone) devices with some consistency, you should use XHTML-MP. If you confidently expect the bulk of your usage to come from smartphone models, go with XHTML Basic (Fling 2009, 173). Brian Fling provides a very helpful "Device Matrix" that lays out the differences with more detail in a way that's understandable to the beginner (2009, 170).

Next, as we've discussed, the number of handheld devices is growing at a rapid rate, and there are a huge number of browsers, so it's very difficult to ascertain for sure how a page is going to look in all of them. We mentioned differences in screen sizes previously; also remember that data download conditions are going to vary considerably, so you need to keep the page sizes small—the recommendation is 20KB or less (W3C 2008). The W3C has a Mobile Web Best Practices document you'll want to familiarize yourself with, particularly section 3.7, "Default Delivery Context," which discusses recommended file types and such. They also have a handy "cheat sheet" that lists what attributes are recognized in XHTML Basic 1.1 (Hazaël-Massieux 2008). As you code, you will also want to strive to employ fluid measures that will adapt to the variously sized displays by using relative units (small, medium, large) or percentages, rather than a fixed layout that would use absolute measures like pixels. As you get more comfortable with CSS, you could explore em measurements, which take a little calculating (Marcotte 2009).

The moral of all this is to do the best you can. There are emulators you can use that will help; enlist your colleagues, friends, and family to try and get a look at how your site displays on as many devices as you can. A word of caution and comfort here—your site is probably going to fail a lot of the checks the first, or second, or fortieth time through. It will probably also inevitably look awful on some mobile browsers. This comes with the territory. You can only do what you can do.

In keeping with our "Start small and iterate, iterate, iterate" philosophy, the key is to get something out there. Once you've tested as best you can, just put something up. You can monitor its use, continue to ask questions, and endlessly update and enhance your site as you go along and as your skills improve. There's always room for getting fancier, prettying up the look and feel, adding features, incorporating location awareness—the sky's the limit. Although we're not recommending this as a best practice, our site-development process started with some off-the-cuff, "we should do that" comments. One day we had a conversation, decided it was a go, and four hours later, it was up and running, with no fanfare, no big announcements, and (almost) no content beyond hours, phone numbers, and a couple of links to our catalog, Google Books, and WorldCatLocal for Mobile. Since then we've improved the underlying code, added features like IM reference, and linked to staff information that we dynamically pull from a database. Not all the new features work on all phones, and we've accepted that.

5.   If you have no IT department, no budget, no CMS, no ability to code…

We won't lie. It's not a pretty scene. There are a few other things you could consider. One interesting option would be to use a transcoding utility. Transcoding is a way of automatically simplifying a "regular" web page to make it more friendly to mobile devices. Free options for this include Google, Mowser, and Skweezer. Since this is accomplished by "prepending" code to a URL, it can be applied on a page-by-page basis. This simply means adding a string of text prior to an existing URL; the most common example you may run across in your daily work would be in proxying URLs to subscription resources to allow access to off-campus users (e.g., the proxy prepend for DePaul University Libraries is http://library.depaul.edu/checkurl.aspx?address= and a full link would look like this: http://library.depaul.edu/checkurl.aspx?address=http://www.springerlink.com/).

It's easier to envision with a concrete example:

Figure 37: ALA Home Page, as viewed on desktop

Figure 38: ALA Home Page, as viewed on iPhone via Google Transcoder

Figure 39: ALA Home Page, as viewed on iPhone via Skweezer

Prepend Codes

http://www.google.com/gwt/n?u=
http://www.skweezer.com/s.aspx?q=

For pages or sites with RSS feeds, which might be the case if you are managing your site using a blog or wiki platform, there are some utilities that will automatically create a mobile site. Again, just to name two: Mippin (free) and Mofuse (paid). An example:

Elmhurst College Library

Figure 40: Elmhurst College A.C. Buehler Library site, desktop

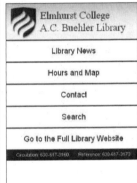

Figure 41: Elmhurst, as viewed on iPhone

Figure 42: Elmhurst College A.C. Buehler Library site, as (minimally) processed by Mippin

Finally, you can avail yourself of a mobile site builder. A couple of examples of free, commercially supported services include MobiSiteGalore and Winksite. Each offers a series of WYSIWYG templates to easily create simple mobile sites; they also host the sites, so you would not have to have access to your library's server environment. One option coming out

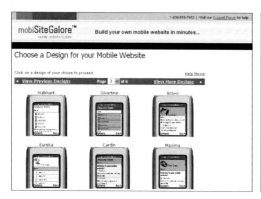

Figure 43: MobiSiteGalore

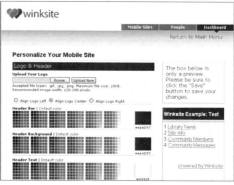

Figure 44: Winksite

of the academic library community which will allow you to quickly create a site optimized for smartphones is Mobile Site Generator. Chad Haefele at University of North Carolina Libraries has created a form-based generator that can be used to create a site structure; once that is done, you would simply need to edit the HTML code to actually add your content. It should be noted that this option will require that you host your own site—that is, that you have the ability to upload files to a server.

To Redirect or Not to Redirect, That is the Question

Once you've got your site running, how are you going to point people to it? You could simply add a link or icon somewhere on each page of your "regular" website, or you could apply a redirect, which will detect the type of device and automatically push users into your mobile site. Griggs, Bridges, and Rempel discuss the various levels of redirection based on device detection (2009):

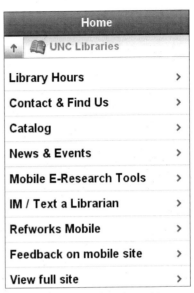

Figure 45: University of North Carolina Libraries mobile site, created using Mobile Site Generator

1.   Different stylesheets
You can have a fancy-pants stylesheet aimed at smartphones, and a more stripped-down stylesheet for the others.

2.   Multiple page versions (requires a server-side script, and some programming knowledge)
The risky thing acknowledged by them, and by everyone who writes about any redirection technique of this kind, is that it relies on the device to tell you what it is through something called the UserAgent, which is "inconsistent, and maintaining a list of possible browsers and operating systems is a constantly moving target." They propose a method that involves identifying desktops and culling them out.

3.   Auto-generate different versions

We've already discussed this in the previous section on transcoding, etc. The jury is definitely out on this one. There are advantages to automatic redirection, and there are liabilities; certainly it's not something that needs to be in place at launch.

Your Mobile Design and Development Toolbox

Like any toolbox, this is just a starter kit. You will need to add specialized implements based on your needs. Any one of the lists of suggested reading could be so much longer. Consider each simply a jumping-off point.

Web Design

- *The Design of Sites: Patterns for Creating Winning Web Sites* (2nd ed.) by Douglas K. van Duyne, James A. Landay & Jason I. Hong
- *Designing Web Usability* by Jakob Nielsen
- *Don't Make Me Think* by Steve Krug
- *The Elements of User Experience: user-centered design for the web* by Jesse James Garrett
- *Getting Real* by 37signals
- *Mobile Design & Development* by Brian Fling
- *Mobile Web Design* by Cameron Moll
- *A Project Guide to UX Design* by Russ Unger and Carolyn Chandler

A List Apart http://www.alistapart.com/
Boxes and Arrows http://www.boxesandarrows.com/
Smashing Magazine http://www.smashingmag.com/

Useful Twitter feeds:
@smashingmag
@reencoded

Standards and Accessibility

- *Designing with Web Standards* (3rd ed) by Jeffrey Zeldman
- *Standards*, World Wide Web Consortium, http://www.w3.org/standards/
- The Web Standards Project, http://www.webstandards.org/
- *Web Standards Solutions: The Markup and Style Handbook* (special ed.) by Dan Cederholm

Some Commonly Used Open-Source CMS

Wordpress: http://codex.wordpress.org/Styling_for_Mobile
[plug-ins: http://wordpress.org/extend/plugins/search.php?q=mobile]

Drupal: http://groups.drupal.org/mobile
[plug-ins: http://drupal.org/project/modules?text=mobile]

Django: http://code.google.com/p/django-bloom/

Mobile Development

- *Mobile Design & Development* by Brian Fling
- *Mobile Web Design* by Cameron Moll
- mobiForge http://mobiforge.com/
- "Mobile Web Best Practices 1.0: Basic Guidelines" by W3C. http://www.w3.org/TR/mobile-bp/
- "Global Authoring Practices for the Mobile Web" by Luca Passani. http://www.passani.it/gap/
- W3C mobileOK Checker http://validator.w3.org/mobile/
- mobiReady—dotMobi compliance & mobileOK checker http://ready.mobi/launch.jsp?locale=en_EN

Mashable: Programming and Web Development Lists, Resources & How-Tos
http://mashable.com/category/programming-lists/

Useful Firefox extensions:
Firebug: http://getfirebug.com/html.html
Web Developer: https://addons.mozilla.org/en-US/firefox/addon/60

Useful Twitter feeds:
@AndroidDev
@iphone_dev

HTML, XHTML and CSS

Everyone has their own ideas about what constitutes the "indispensible" guide in this area. Here are just a few options you might consider.

- *CSS: The Missing Manual* (2nd ed.) by David Sawyer MacFarland
- *CSS Mastery: Advanced Web Standards Solutions* (2nd ed.) by Andy Budd, Cameron Moll, and Simon Collison
- *HTML, XHTML, and CSS* (6th ed.) (Visual Quickstart Guide) by Elizabeth Castro
- *HTML & XHTML Pocket Reference and Web Design in a Nutshell: A Desktop Quick Reference* by Jennifer Niederst-Robbins
- A few other authors to check into: Dan Cederholm, Molly E. Holzschlag, Eric Meyer, Charles Wyke-Smith

A Few Handy Code Snippets

These may be helpful in starting your pages (this is intended strictly as a skeleton for a mobile site):

XHTML Mobile 1.1

```
<?xml version='1.0' encoding='UTF-8'?>
<!DOCTYPE html PUBLIC '-//WAPFORUM//DTD XHTML Mobile 1.1//EN'
'http://www.openmobilealliance.org/tech/DTD/xhtml-mobile11.dtd'>
```

```
<html xmlns="http://www.w3.org/1999/xhtml">
<head>
<title>Site Name</title>
<meta http-equiv="content-type" content="application/xhtml+xml" />
<meta http-equiv="cache-control" content="max-age=600" />
<link rel="stylesheet" type="text/css" href="mobile.css"/>
</head>
<body>
<H1>My Page Title</H1>
<p>All of my exciting content. Also, contact us!</p>
<p><a href="tel:2125551211">call us at 1(212)555-1211</a></p>
<p><a href="sms:12125551212">Send SMS to 1(212)555-1212</a></p>
</p>
</body>
</html>
```

XHTML Basic 1.1

```
<?xml version="1.0" encoding="utf-8"?>
<!DOCTYPE html PUBLIC "-//W3C//DTD XHTML Basic 1.1//EN"
"http://www.w3.org/TR/xhtml-basic/xhtml-basic11.dtd">
<html xmlns="http://www.w3.org/1999/xhtml" xml:lang="en">
<head>
<title>Site Name</title>
<meta http-equiv="content-type" content="application/xhtml+xml" />
<meta http-equiv="cache-control" content="max-age=600" />
<link rel="stylesheet" type="text/css" href="mobile.css"/>
</head>
<body>
```

```
<H1>My Page Title</H1>
<p>All of my exciting content. Also, contact us!</p>
<p><a href="tel:2125551211">call us at 1(212)555-1211</a></p>
<p><a href="sms:12125551212">Send SMS to 1(212)555-1212</a></p>
</p>
</body>
</html>
```

Chapter 5
Ready Set Go: Launching Your Mobile Site

Though your mobile site may start small—perhaps less than ten pages altogether—it still deserves the same careful roll-out you'd plan if launching a full-scale website for your library. Notwithstanding our confession in chapter 4 about our necessitated on-the-fly development process (having had the experience, we can tell you it's definitely not a best practice), you only have one chance to wow your users. And once your mobile site is out there, trust us: your users will find it, ready or not.

A good rule of thumb is to always launch a new service on a Monday (and your mobile site is absolutely a new service). If you launch on a Monday, you'll have all week to work out the kinks that come with any type of roll-out: wonky links, platform-specific display issues, user consternation at seeing something new, etc. Never launch on a Friday, when disaster could strike over the weekend and you and the rest of the people who would need to know will probably be unavailable.

A successful launch will involve both internal and external user testing, staff training, and smart marketing. This will usually take place in three stages: an internal launch, a soft launch, and finally, achieving critical mass.

Talking the Talk

Beta: a test for a computer product prior to commercial release. Beta testing is the last stage of testing, and normally involves sending the product outside the company for real-world exposure.

Focus group: a small group of people (usually 5-8) who react to ideas and designs presented to them; good for quickly getting a sampling of users' opinions and feelings about things.

Internal launch: releasing your product to users within your organization to test functionality and encourage high levels of of awareness and commitment to the new product.

Silent launch: making a product live and available to the public without an announcement or promotional items.

Soft launch: releasing your product to a limited audience, accompanied by little or no marketing, with the intent of making revisions based on user feedback before making it generally available as a hard launch.

Usability testing: one user at a time is shown something (like a website), and asked to figure out what it is or try to use it to do a typical task.

Internal Launch and Staff Advocacy

An internal launch will mean generating excitement within your organization and asking coworkers to test the site. This is a low-risk, low-cost approach to doing some initial user testing before exposing your site to the public. During the internal launch phase, your goal should be to get as many users as possible to try to "break" the site. They can not only give you feedback on their overall impressions of your site's aesthetics but also check for grammar errors, spelling mistakes, broken links, improper display of graphics, unusual lag time when new pages are loading, and any 400 error codes that might pop up during testing. These are all things that you won't be able to take back once your first external user encounters them; if your users are like ours, the rule of "once bitten, twice shy" applies: most students are reluctant to try a system a second time that acted improperly the first time around.

The other important step the internal launch phase accomplishes is ensuring all members of your staff feel involved in the project of creating a mobile web presence, which goes a long way toward greasing the buy-in wheel. You want your coworkers to be excited about your institution taking this next step in web design, and you want them to be able to talk knowledgeably (and enthusiastically!) about your mobile site if they encounter a user with a question; every interaction is a marketing opportunity.

Structured correctly, your internal launch can be extremely productive. Initially, you may want to preface your launch with an open invitation for coworkers to freely explore the site and test out the functionality as if they were the end user, making sure you accompany all communications with a call for feedback. Once they are comfortable with the site, and you have made any necessary edits from the first round of comments, develop a task list to distribute. Ask coworkers to test everything from links (did the link do what you thought it would do?) to forms (was the data captured completely and did it end up in the correct person's inbox?) to page comprehensiveness (was there anything on the page that you expected to see that you did not?) to display (did anything look out of place?). Testing guidelines could also include ease of navigation, consistency throughout the site, and speaking the user's language.

Soft Launch

A soft launch involves releasing your site to small groups of external users for a few rounds of user testing. A soft launch is also sometimes referred to as beta testing; labeling your website as a beta product means users will be more tolerant when everything isn't working perfectly. Probably the most widely known form of user testing is a usability test, in which users are observed while completing a series of typical tasks (U.S. Department of Health and Human Services n.d.). If you haven't already read Steve Krug's *Don't Make Me Think*, pick up a copy (2006). Immediately. Krug's everyman approach to usability testing explains it in such a way that literally anyone can complete a round or two of effective user testing, and he covers everything from why you absolutely should do it to recruitment of participants to acting on the results. He's also highly entertaining, so definitely read his book for yourself. Based on the usability-testing foundation that Krug lays down, the following is a quick rundown on why it would be shortsighted not to explore user testing as an option before making your site live.

Anyone can conduct a round of user testing; all you need is a modicum of people skills, some patience, and a thick skin. Luckily, these are usually skills public-service librarians already possess. As the front person for the usability study, you'll want to put the participant at ease, which involves finding out a little about the participant, stressing that he can stop the test at anytime he feels uncomfortable, and assuring him that he will not hurt your feelings, no matter what he thinks about the website. And while you might put a time limit on each task you are asking the participant to complete, you want to ensure he feels comfortable exploring the website and taking his time to weigh his options when presented with a new task. This is where you'll need to fight the urge not to help, which goes against the very nature of every librarian. Lastly, if a participant struggles with a task, fails it completely, or makes a negative comment regarding the site, don't take it personally. Remember, user testing can only help improve the site, resulting in a more positive user experience once the site is live, and user satisfaction is what we are striving toward.

And you don't need a fancy user-testing lab and high-tech equipment to successfully complete a round of user testing. Really, all you'll need is an empty office or conference room, a computer, and a way of recording the user-testing sessions. Last spring, we completed two rounds of user testing using this very low-tech method. In both rounds, we used a coworker's office, which was located in a quiet corner of the library and was large enough to feel comfortable. In the first round, we set up a laptop with screen-capture software installed on it and placed an audio recording device behind the laptop (so the participant wouldn't be constantly aware of the recorder). These were all things we already owned. We had two librarians scheduled for each session, one to administer the session, and one to observe. It was that easy.

For the second round, we took the plunge and upgraded to a software application called Morae specifically designed for conducting usability tests. It's akin to a fancier version of our screen-capture software that offers features like automatically recording important statistics (mouse clicks, time elapsed per task) within the individual sessions, which later saved us hours of coding the data by hand. You can test your mobile site on a desktop computer using different browsers set to a smaller size to mimic a device, and this might be an effective method to test specific pieces of your site, but actual conditions will be quite different—using a mouse instead of a touch screen is a big difference! Also, as we've mentioned before, it's hard to know what to expect: there were over 40 known browsers in play for mobile devices in 2008 (Kroski 2008a). You might find it helpful to ask participants to bring their own web-enabled phones or smartphones to the sessions to actually see your site in action on a variety of devices.

Our low-tech methods are also low cost; user testing doesn't have to be expensive. As mentioned, we took advantage of equipment and space we already owned to keep the cost reasonable. We also went the bargain-basement route with recruitment, so the biggest costs were staff time, paper, and color copies for fliers and posters to put up around campus. Our biggest non-staff expense was the incentive of $25 gift cards to our campus bookstore. In recruiting college students, it doesn't take a lot of money to pique their interest! Just mentioning compensation on a sign was enough to get most students to amble over to talk to us.

Finally, it's never too late for user testing (that said, you can never test too early). There certainly comes a time when you become too close to your site, too saturated in it, to see it objectively. Just having one external user give it a spin can give you fresh insight into the design and functionality of your site. However, imagine how much insight you'll gain from having multiple users test it during various stages in the development process.

Two important notes: because usability tests typically do not involve a large sample of users, their results are most useful when a series of tests are conducted and compared. That is to say, conducting a test, making small changes, and conducting a second test of the same tasks will allow you to assess whether the changes constituted improvement. Lastly, if you plan on ever using data collected during user testing for anything other than internal documents, you must follow your institution's human-subject research policies and file the appropriate paperwork before beginning testing.

Achieving Critical Mass through Marketing

Once your site is ready to go live, you'll want to spend some time developing a marketing strategy. The worst thing that could happen at this stage would be that no one knows about your mobile site after you've invested so much time and energy in it.

A sound marketing plan can also reach those users who might not go fishing around on their mobile devices, hoping to land on a mobile site for your library. While there are other techniques we've mentioned to drive traffic to your site—for example, within the first month of the silent launch of our mobile site, we had significant enough traffic from a variety of smartphones that we took the plunge and put a redirect in place to detect and funnel traffic automatically to our mobile site—those cannot replace the value of a well-thought-out marketing effort.

At the foundation of any marketing plan are the four Ps: product, place, price, and promotion (Fisher and Pride 2006, 24). While we are nonprofit organizations, libraries shouldn't

Mobile Marketing Tips

Get the word out as widely as you can, in a variety of mediums. Advertise on your blog, Twitter feed, Facebook page, library homepage, and through traditional (print) methods, such as posters and fliers displayed around your library and campus, an article in your student newspaper, or town newsletter.

Create an incentive for users to try out your mobile site. When we first launched our Facebook page, we offered a goodie bag as a gift to the first users who signed up as "fans" of the library; as a result, we had close to 50 fans within the first month. It doesn't need to be expensive. Our bags included library-branded pencils, notepads, and some candy.

Add your mobile site to directories and aggregators (like the M-Libraries wiki, or find out if your school or town is maintaining its own directory of mobile friendly sites).

[Adapted from Kroski 2008b]

shy away from employing techniques that have been extremely successful in the for-profit world. Your mobile site is a product that has value and can assist users with their information needs: you need to convince them of this. Having a marketing team can be invaluable as you brainstorm ideas about the Product you are offering (your mobile site), the Place (anywhere, anytime), the Price (it's free), and the Promotional techniques employed to ensure its use.

The last P, Promotion, is always the most gratifying. This is when you finally get to send your site out into the world. The most common marketing tools and techniques include advertising, brochures, endorsements, contests, direct mail, exhibits, posters, and videos, so don't feel limited to just posters and fliers. As Alison Circle and Kerry Bierman put it, "The days when marketing was thought to be posters and fliers is over. In today's world, marketing is at the core of every transaction, from checkout and customer interaction to story time and our buildings themselves" (2009, 32). The importance of involving every staff member in the production and launch of your mobile site cannot be emphasized enough.

This ensures a successful patron-staff interaction every time a question arises. These days, as we've often experienced in our own lives, marketing as a professional field is increasingly trending towards a focus on one-to-one customer relations.

Be sure to brand your site, too. Your mobile site should clearly be an extension of your library, which involves making it look and feel akin to your other products and services. And don't be afraid to try out a new marketing tool. This is the time to be creative. There are many low-cost ways to increase the visibility of your mobile site; just one idea might be to embed a short YouTube video on your main website giving a demo of how to access and use the main features of your mobile site.

Fig. 46: Orange County (FL) Library System: YouTube video advertising mobile site

Fig. 47: Orange County (FL) Library System: What's New page highlighting services

The easiest place to start developing your marketing plan might be by thinking about who your target audience should be. Even if the majority of your population falls within the traditional age range for college students (incidentally, also the group most likely to readily adopt mobile technologies), you might still choose to focus on a particular segment of the group, such as freshmen, to develop your marketing strategy. What are their likes and dislikes? What language should you use to reach them? How does your mobile site benefit them specifically? Then develop your product/service description. This product description will live in the introduction of your marketing plan and should clearly explain to any reader the product and its value.

There are seven main sections that should be present in any solid marketing plan, and many more that you can include (Fisher 2006, 27):

- Introduction/executive summary
- Service/product description
- Target market description

- Marketing goals and objectives
- Strategies and tactics
- Implementation
- Summary/Conclusion

By design, marketing plans reflect short-term strategies and should be revisited every year, if not more frequently.

There are various templates and guides you can use to produce a fully fleshed-out marketing plan. These include more library-focused materials provided by ACRL and ALA, to the classic text *Marketing Plans: How to Prepare Them, How to Use Them* by Malcolm McDonald. You can also take advantage of online resources marketing professionals use, such as marketingprofs.com and marketingjournal.blogspot.com. We've included a sample plan for you, together with questions to help you get started writing your own.

Sample Marketing Plan

Introduction

Questions to answer in this section:

- What is the new product or service?
- Why is it important?
- What is the background or timeline for this project?
- What need will it fill?

Noting a trend in students using mobile devices to look up items in the library's catalog, in spring 2009 the John Q. McFakeguy Library decided to develop a website optimized for viewing on a mobile device. A scaled-down version of the current library website, the mobile site is limited in its functionality and is designed for quick, on-the-go tasks, such as looking up library hours. Construction was completed by June, and over the summer, several rounds of internal testing and user testing were completed to ensure a solid design and scope. Our mobile site is equipped to handle common tasks, and gives users an easy access point to look up library hours, contact information, locations, news and events, and a quick title search, as well as leveraging mobile versions of certain vendor resources, like WorldCat Mobile. Users can access the mobile version from any mobile device with an internet connection from anywhere in the world. The hard launch is set for the beginning of fall semester, to coincide with the students returning to campus, and the freshman population is the target market segment.

Demographics suggest that the highest number of mobile technology adopters fall within the 18 to 28 age range, and we feel that our mobile site and services will serve as a solid promotional technique to encourage new students to use library resources.

Current Situation

Questions to answer in this section:

- What is the state of the product?
- Are there any technical issues that need to be dealt with?
- Has the product been field tested?
- How do you envision your users interacting with the product?

The mobile site can be accessed anywhere in the world from any device with an internet connection. Currently, the site is maintained using the existing content management system, and the only expenses have been in staff time and talent. Resources available to students are freely offered, although some may require authentication. Most current mobile web users are already used to accessing mobile versions of popular websites, such as Google, CNN, Amazon, Wikipedia, and Facebook, so the learning curve for our mobile site should be quite low. Based on initial results of user testing, our site is highly intuitive and user-friendly.

Marketing Goals

Questions to answer in this section:

- What do you hope to get from your marketing campaign?
- How will you evaluate the results?

By the end of fall semester, we hope to see an increase in awareness of our mobile site, not just by the incoming freshmen population, but across student groups. Usage will be tracked using Google Analytics, and we expect to find significant traffic on our mobile site before we break for winter intersession. A feedback form will be included on our site, and pressing changes will be made on an as-needed basis, with a more formal evaluation scheduled for after finals.

Marketing Strategies

Questions to answer in this section:
- What marketing techniques will you employ?
- Will any user education be necessary?
- What media outlets, if any, will you utilize?

While we feel that our strongest marketing tactic will be word of mouth (both student to student, and librarian to student), we will employ techniques such as a blog entry announcing the launch of the mobile site, as well as more traditional approaches: posters displayed in the library and fliers posted around campus, especially in the dorms. All promotional items will contain the mobile site URL and the tagline: "On the move? Now, so are we!" and will be branded with the library's name and logo.

Short, instructional videos will also be produced and linked from our homepage, explaining how to use the mobile site, and highlighting the main features to which we want to direct student attention. Since we are targeting students for this resource, an ad will also be placed in the student newspaper to run once a month for the three full months of the fall semester. Train-the-trainer sessions will also be held internally to ensure all staff members are aware of the mobile site, can successfully interact with the site, and have some exposure to mobile devices.

Implementation

Questions to answer in this section:
- How will you put your plan into action?

- Will you need to seek approval for any part of your marketing plan?
- Will you need to work with any other departments or groups within your organization?
- What is your timeline?
- How will you stay on-task?

The marketing committee will work on finalizing the design of the posters and fliers, and will seek final comments and approval at the August Public Services meeting. This design could be expanded to other promotional items such as moo (mini-promotional) cards or bookmarks. The posters and fliers will then be distributed before the fall semester begins in September. The marketing committee will also draft the announcement for the library's blog, for posting during the second week of classes to allow freshmen a "settling in" period. Marketing will also partner with the student newspaper to develop the ad for the mobile site. The instructional-design team has volunteered to work with the marketing committee on creating the instructional videos, which should go live during the first week of classes.

A detailed timeline and task list will be developed and monitored closely by the marketing committee, with a point person assigned to each task. A progress report will be delivered during the bimonthly marketing meetings.

Summary/Conclusion

We feel that launching our mobile site is an important step for our library to take. Not only are we utilizing a popular technology, but we are also staying relevant with a changing user population. Our mobile site will be convenient for our users, and is accessible from devices currently in use across campus. This resource will also help to forge a relationship with a new group of users, and introduce them to additional resources provided by the library. We feel that our current marketing strategy will ensure awareness and use of our mobile site, and we will monitor usage throughout the fall semester.

Further Reading
User Testing and Usability
- *Designing Web Usability* by Jakob Nielsen
- *Don't Make Me Think* by Steve Krug
- *Handbook of Usability Testing: How to Plan, Design, and Conduct Effective Tests* by Jeffrey Rubin
- *Observing the User Experience: A Practitioner's Guide to User Research* by Mike Kuniavsky
- *Task-Centered User Interface Design: A Practical Introduction* by Clayton Lewis and John Rieman (freely available at: http://www.hcibib.org/tcuid/)
- *The Design of Everyday Things* by Don Norman (for a broader based view on design in the world)
- A List Apart http://www.alistapart.com/
- Boxes and Arrows http://www.boxesandarrows.com/
- Jakob Nielsen http://useit.com/
- Jared Spool, User Interface Engineering http://www.uie.com/

Marketing
- *Blueprint for Your Library Marketing Plan: A Guide to Help You Survive and Thrive* by Patricia H. Fisher and Marseille M. Pride.
- *Marketing Plans: How to Prepare Them, How to Use Them*, by Malcolm McDonald.
- http://www.marketingprofs.com
- http://marketingjournal.blogspot.com

Chapter 6
Tending Your Garden: Evolution and Iteration

At this point, you may decide this mobile racket is not for you. While this may seem like counterproductive advice upon reaching this chapter, you also need to make the right decision for your library: if you don't have the time, resources, or inclination to keep up a mobile site, sometimes the best course of action is to not launch a mobile site. If this is the path you choose to take, there's still good news. If you have a website for your library, you are already a part of the mobile web: "Between advanced devices such as the iPhone, innovative browsers such as Opera, and automatic transcoding by major search engines, your site might not look all that bad with no extra effort on your part" (Kroski 2008, 40). We strongly recommend testing your library's website on a mobile device to get an idea of how it will display on a Blackberry, for example, and urge you not to let the mobile-web conversation die at your organization. Even if now is not the time to launch a mobile site, you may reach a different decision six months down the road.

If you do decide to move forward with planning, building, and launching a mobile site, then congratulations! This is going to be fun! Just like with any web presence, when it comes to your mobile site, static is a bad word. However, as we've emphasized throughout this book, be careful to only give your mobile users features that will be useful to them. Practically speaking, your mobile users will most likely not launch any heavy-duty research projects from their devices, but they might look up a word in the OED to settle a bet at a party. And they might appreciate an interactive map of your location. There are a lot of smart people throwing serious money into the mobile web, which means you can benefit from their work by hooking up to services like WorldCat Local or Google Maps or the mobile interface offered by EBSCOhost. There's plenty you can do to keep your mobile site fresh and innovative without doing a lot of the heavy lifting yourself. But it does mean you will need to devote some time and effort to keeping abreast of new developments in mobile technologies.

Keeping Up With the Joneses

Pew Internet and American Life Project
http://www.pewinternet.org

Library Technology Reports
http://www.alatechsource.org/ltr/index

M-Libraries
http://www.libsuccess.org/index.php?title=M-Libraries

The Horizon Report
http://www.educause.edu

Another easy way to ensure you're providing your users with useful information and service points on your mobile site is to provide a place for user feedback. Make sure you include contacts, e-mail addresses, and phone numbers on materials that publicize your site, and actively solicit comments and suggestions using a feedback form on your site. Carefully consider what your users are saying in their responses. You don't have to act on every suggestion that comes in, but at least deliberate on what your users are telling you and where they are coming from. Think about sorting your feedback into at least the following three categories: what has to be changed? What can easily

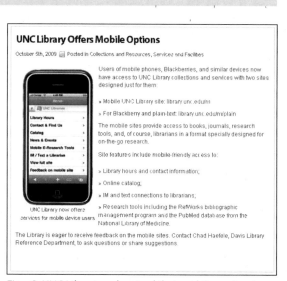

UNC Library Offers Mobile Options

October 5th, 2009 Posted in Collections and Resources, Services and Facilities

Users of mobile phones, Blackberries, and similar devices now have access to UNC Library collections and services with two sites designed just for them:

» Mobile UNC Library site: library.unc.edu/m
» For Blackberry and plain-text: library.unc.edu/m/plain

The mobile sites provide access to books, journals, research tools, and, of course, librarians in a format specially designed for on-the-go research.

Site features include mobile-friendly access to:

» Library hours and contact information;
» Online catalog;
» IM and text connections to librarians;
» Research tools including the RefWorks bibliographic management program and the PubMed database from the National Library of Medicine.

UNC Library now offers services for mobile device users

The Library is eager to receive feedback on the mobile sites. Contact Chad Haefele, Davis Library Reference Department, to ask questions or share suggestions.

Fig. 48: UNC Libraries advertised their mobile services in their blog

be changed? What are wishlist items for future iterations? Your feedback form could be your most effective, least expensive method of identifying and developing positive upgrades to your site.

As we mentioned in chapter 4, user-centered, results-based interaction web design is important. This school of thought strives to understand, include, and satisfy users, rather than just push out information with a take-it-or-leave-it attitude. Interaction design focuses on user goals and ultimately tries to create a website that is "usable, meaning easy to learn, useful, and an enjoyable user experience" (Weiss 2008, 43). What's great about this approach to web design is that it mirrors the positive customer-service atmosphere already present in a library setting; we're already committed to giving our users what they want, and we can ensure this way of thinking also manifests itself in our websites. Meredith Weiss does a very effective job of summarizing interaction design and its impact within higher-education settings. But what struck us the most in reading her article was her proposed REACH framework for developing a website:

- Research (Marketing)
- Experiment
- Assess and Analyze
- Construct (Interactive Design)
- Honor Your Findings to Improve Results

As you can imagine, it was the "H" that captured our attention. Too often, we develop a website full of shiny new content, only to have it languish and mold on the interwebs. We provide feedback forms, and then rarely bother to check the inbox. And we throw launch parties, only to forget to attend to the care and feeding of our sites. So our reiterated message to you, dear reader, is to honor your findings to improve results. Weiss recommends six ways to achieve the results you want (46):

- Review analyses and evaluations
- Use data gathered to achieve organizational goals
- Use data gathered to improve user experiences
- Review and update website goals
- Refine the website
- Continually measure the website's impact and REACH toward building institutional value

One tip to keeping your site agile in a rapidly changing technology environment is to make it device independent. We began this discussion in chapter 4, and it bears repeating here. This will make updates easier in the future, since you can worry less about what your users have to access your mobile websites and more about staying aligned with web standards, which change less frequently and less dramatically than the latest smartphone models.

A few other easy methods for evaluating the effectiveness of your site include monitoring Google Analytics to see what is actually being used on your site, informally polling your student workers for their thoughts on your mobile site, and holding later rounds of user testing, perhaps six months or a year after you've launched your site.

Fortunately, and unfortunately, the possibilities of services and information delivered in a mobile environment are limitless. While this is an exciting time, it also means that real time and energy will need to be devoted to this new trend in technology. Even in the short run, there are still many things libraries could be offering for the mobile web, such as developing brief guides on specific topics pointing to resources that are designed to be viewed on a mobile device, or taking advantage of vendor offerings of mobile-optimized dictionaries and reference sources. What you'll have to decide is what initiatives make the most sense for your organization and align most closely with your strategic plans.

Whether you are an academic library, or a public or special library, we now turn to the importance of getting involved in the larger mobile-web discussion going on in your community. Joan Lippincott, in her ARL report "Mobile Technologies, Mobile Users," urges libraries to get into the mobile web before it's too late.

> "If the library is not at the table, will other campus units make decisions that result in incompatibility with equipment and content purchased and licensed by the library?… As with most technology developments, this one is fast-moving. This is not a time to sit on the sidelines as other campus units are developing services for mobile users and licensing content for mobile devices. Academic libraries should make conscious choices about what they want to offer in this arena and act accordingly." (Lippincott 2008, 4).

This advice applies to campus IT, or the mayor's office, or citywide planning commissions. Make sure you're at the table. Not only do you need to give your library a voice, but you need to know broadly what your community is planning related to technology.

In academic libraries, be sure to talk regularly with your IT department. Currently, many institutions are exploring the possibilities of offering a whole host of mobile services to students, including delivery of exam results, admission status, course registration, and student account access. You might see where the library can benefit from piggybacking onto these initiatives, and use similar systems to provide SMS overdue alerts or renewals. We recently learned that our institution had formed a social-media group, so we lobbied to be part of the initiative. Initially, larger departments within the university didn't know we were on Facebook and Twitter and actively using other social media. After making connections, we now have a library representative who attends the monthly meetings, and we're better positioned because of it.

For those in public libraries, the same strategy applies. Get a representative involved in town council meetings, and find out if your city is participating in a broadband initiative program via the economic stimulus plan. If your community is launching an emergency-alert system that will communicate with cell phones, you may be able to use the same system to send out SMS notices to your patrons who choose to opt in, or develop a geocaching game for your community to participate in using their GPS-enabled handheld devices to encourage awareness of the features available on mobile devices.

Trend Tracking

While technologies are constantly evolving, and there will also surely be the next Next Big Thing on the not-too-distant horizon, there are ways to build from your mobile-web momentum in your organization. Consider the development of a mobile site as a launching point, rather than a discrete, isolated project. The mobile web is not going away, and all signs point to it only increasing in popularity. After all, according to the Pew Internet and American Life Project report "Mobile Access to Data and Information," over half of young adults surveyed said it would be very hard to give up their cell phones, and 62 percent of all Americans have used a cell phone or PDA for a nonvoice data application, such as texting or e-mailing (Horrigan 2008, 1). As we move from e-learning to m-learning, libraries still very much have a vital role in our communities; there is no price tag on creativity.

Jim Hahn goes so far as to suggest that the library serve as a lab space for m-learning services and technologies, and it seems like a natural fit (2008, 284). Libraries have always been a safe space for experiments in learning. It might seem like a scene from a sci-fi movie to imagine a student walking through a library scanning book titles with her mobile phone and pulling up real-time reviews, but it's not that far off. Hahn also persuasively argues that embracing m-learning technologies could have a positive effect on lessening library/research anxiety, and that taking full advantage of the hardware and software already standard on most mobile devices would mean creating new services and features that do not have an analog equivalent. This would mean transforming the library into something new, that "the existence of the library in mobile digital form will be something altogether unequal to the physical realm. The m-library is not truly the library on a portable device but rather a new unequivocal resource that can reshape scholarship, study, research, and librarianship" (Hahn 2008, 284). It's an exciting time to be in the information business.

But you don't have to transform your library overnight into a hub of mobile services! Start small, for example, with a mobile site. It's enough, right now, just to be in the mobile game, and let other services and resources follow as a natural extension, once you judge that your coworkers, your users, and your library are ready. And how do you know what's coming down the pike? A few accessible resources that we rely on are technorati.com, wired.com, Wired Campus from the *Chronicle of Higher Education*, and certainly not least, reading and talking to the many smart librarians out there!

That's All, Folks!

So, now what? We've said all along that you should think of this book as your starter kit, your getting-started guide, your basic toolbox. When it comes down to it, you have to do what's right for your organization in the current time and space in which you exist. You might find that you do not complete all of the steps outlined in this book in a formalized way, but instead have water-cooler discussions with colleagues about whether or not your users are ready for mobile. Another point we'd like to emphasize is that launching a mobile site does not have to be a long, torturous process; you are not stuck with whatever you put up as your first iteration. Our mobile site has already gone through several versions in the few short months it has been alive. And with each pass, it improves and becomes more functional and more user friendly.

When we started our own experimentation with a mobile site, we wish someone had handed us a guide like this. It certainly would have saved us considerable gnashing of teeth and moments when we thought we had brought the entire system down (nothing can be proven!). And if we can save you some blood, sweat, and tears, then our work here is done; after all, forewarned is forearmed. We'd be delighted to hear of your successes—e-mail your mobile URLs to us at mobilelibraries@gmail.com.

References

37signals (Firm). 2006. *Getting real: The smarter, faster, easier way to build a successful web application.* 1st ed. Chicago, IL: 37signals.

Anderson, Janna, and Lee Rainie. 2008. *Future of the Internet III: How the experts see it.* Pew Internet & American Life Project, December 14. http://pewresearch.org/pubs/1053/future-of-the-internet-iii-how-the-experts-see-it.

Andrew, Paul. 2009. A guide to Google analytics and useful tools. *Smashing Magazine.* July 16. http://www.smashingmagazine.com/2009/07/16/a-guide-to-google-analytics-and-useful-tools/.

Ballard, B. 2007. *Designing the mobile user experience.* Chichester, England: Wiley.

Barba, Shelley, and Joy Perrin. 2009. Great expectations: How digital project planning fosters collaboration between academic libraries and external entities. *Texas Library Journal* 85, no. 2 (Summer 2009): 56-59.

Basso, Monica. 2009. Social trends are influencing the adoption of mobile and web technology. Gartner Inc., August 10.

Cervone, H. Frank. 2007. Standard methodology in digital library project management. *OCLC Systems & Services* 23, no. 1: 30-34. doi:10.1108/10650750710720748.

———. 2008. Good project managers are "cluefull" rather than clueless. *OCLC Systems & Services* 24, no. 4: 199-203. doi: 10.1108/10650750810914201.

Circle, Alison, and Kerry Bierman. 2009. The House brand. *Library Journal* 134, no. 11 (June 15): 32-35.

CoCreatr. 2008. *1. find QR code* Photograph. January 22. Flickr! http://www.flickr.com/photos/cocreatr/2211459923/.

Dempsey, Lorcan. 2009. Always on: Libraries in a world of permanent connectivity. *First Monday* 14, no. 1. http://firstmonday.org/htbin/cgiwrap/bin/ojs/index.php/fm/article/view/2291/2070.

dotMobi. Mowser—Mobilizing the web. http://www.mowser.com/.

Dudden, Rosalind F. 2007. *Using Benchmarking, Needs Assessment, Quality Improvement, Outcome Measurement, and Library Standards: A How-to-Do-It Manual with CD-ROM.* New York: Neal-Schuman Publishers.

Encyclopædia Britannica, Inc (Firm). Britannica Mobile—iPhone Edition. http://i.eb.com/.

Enslin, Rob. 2009. *SketchBook Mobile Express.* September 21. Flickr! http://www.flickr.com/photos/doos/3942284566/.

Fifarek, Aimee. 2007. Blazing the tech trail: tips for implementing new technologies in a risk-adverse environment. *Library Hi Tech News* 24, no. 1 (January): 21.

Fisher, Patricia H. 2006. *Blueprint for your library marketing plan: A guide to help you survive and thrive.* Chicago: American Library Association.

Fling, Brian. 2009. *Mobile design and development: Practical concepts and techniques for creating mobile sites and web apps.* 1st ed. O'Reilly Media.

Garrett, Jesse James. 2002. *The elements of user experience: User-centered design for the web.* 1st ed. Indianapolis, Ind: New Riders.

Giesecke, Joan, and Beth McNeil. 2005. Project management. In *Fundamentals of library supervision,* 146-152. Chicago: ALA Editions, January.

Greenwalt, Toby. 2009. Text messages from your library! *The Radar.* February 13. http://blogs.skok-ielibrary.info/radar/2009/02/13/sms/.

Griggs, Kim, Laurie M. Bridges, and Hannah Gascho Rempel. 2009. library/mobile: Tips on designing and developing mobile web sites. *The Code4Lib Journal,* no. 8 (September 21). http://journal.code4lib.org/articles/2055.

Haefele, Chad. 2010. Mobile Site Generator. Hidden Peanuts. February 9. http://www.hiddenpeanuts.com/archives/2010/02/09/mobile-site-generator/

Hahn, Jim. 2008. Mobile learning for the twenty-first century librarian. *Reference Services Review* 36, no. 3: 272-288.

Haynes, Tom. 2008. The Three-E strategy for overcoming resistance to technological change. *EDUCAUSE Quarterly* 31, no. 4 (December): 67-69. http://net.educause.edu/ir/library/pdf/EQM08411.pdf

Hazaël-Massieux, Dominique. 2008. XHTML Basic 1.1 Cheat Sheet. March 28. http://www.w3.org/2007/07/xhtml-basic-ref.html.

Horowitz, Lisa. 2009. Assessing library services: A practical guide for the nonexpert. *Library Leadership & Management* 23, no. 4: 183-203.

Horrigan, John. 2008. *Mobile Access to Data and Information.* Pew Internet & American Life Project, March 5. http://www.pewinternet.org/Reports/2008/Mobile-Access-to-Data-and-Information.aspx.

———. 2009. *The Mobile Difference.* Pew Internet & American Life Project, March 25. http://www.pewinternet.org/Reports/2009/5-The-Mobile-Difference—Typology.aspx?r=1.

Kaushik, Avinash. 2009. Internal site search analysis: Simple, effective, life altering! *A List Apart*. 9. http://www.alistapart.com/articles/internal-site-search-analysis-simple-effective-life-altering/.

Kroski, Ellyssa. 2008a. What is the mobile web? *On the move with the mobile web: Libraries and mobile technologies*. Library Technology Reports 44, no. 5 (July): 5-9.

———. 2008b. How to create a mobile experience. *On the move with the mobile web: Libraries and mobile technologies*. Library Technology Reports 44, no. 5 (July): 39-42.

Krug, Steve. 2006. *Don't make me think!: A common sense approach to web usability*. 2nd ed. Berkeley, Calif: New Riders.

Kuniavsky, Mike. 2003. *Observing the user experience: A practitioner's guide to user research*. San Francisco, CA: Morgan Kaufmann Publishers.

Lippincott, Joan K. 2008. Mobile technologies, mobile users: Implications for academic libraries. *ARL: A Bimonthly Report on Research Library Issues & Actions*, no. 261 (December): 1-4.

Lorcan Dempsey. 2008. Foreword. In *M-Libraries: Libraries on the move to provide virtual access*. London: Facet.

Mak, Venia. 2009. What mobile services does your library offer? *LibraryConnect* 7, no. 4 (November): 8.

Marcotte, Ethan. 2009. Fluid grids. *A List Apart*, March 9. http://www.alistapart.com/articles/fluidgrids.

Marill, Jennifer L., and Marcella Lesher. 2007. Mile high to ground level: Getting projects organized and completed. *Serials Librarian* 52, no. 3/4 (June): 317-322.

Moll, Cameron. 2007. *Mobile web design*. Salt Lake City: Cameron Moll.

MOO, Inc. MOO. http://us.moo.com/en/.

Nielsen, Jakob. 2009. Mobile usability. *Jakob Nielsen's Alertbox*. July 20. http://www.useit.com/alertbox/mobile-usability.html.

Papp, Meredith. 2009. Introducing Google Analytics for mobile apps. *Official Google Mobile Blog*. November 3. http://googlemobile.blogspot.com/2009/11/introducing-google-analytics-for-mobile.html.

Parr, Ben. 2009. Mobile Web Is Taking Over the World (and Other Internet Trends). October 20. http://mashable.com/2009/10/20/mobile-web-presentation/.

Passani, Luca. 2009. Global authoring practices for the mobile web. September. http://www.passani.it/gap/.

Pin, Wan Wee, Liau Yi Chin, and Chua Lay Lian. 2009. The library in your pocket: Making the library truly accessible anytime, anywhere. *LibraryConnect* 7, no. 4 (November): 6.

Projeto Sticker Map. 2008. *QR code*. October 12. Flickr! http://www.flickr.com/photos/sticker-map/2935588694/.

Reese Brown, Dakota. 2009. Four key principles of mobile user experience design: Academic insights tested in the real world. *Boxes and Arrows*, November 19. http://www.boxesandarrows.com/view/four-key-principles.

Regents of the University of Michigan. 2009. Proxy server bookmarklet. July 6. http://www.lib.umich.edu/mlibrary-labs/proxy-server-bookmarklet.

Reisinger, Don. Touch-screen phone use soars, iPhone on top. *CNET*. http://news.cnet.com/8301-17938_105-10389847-1.html.

Royse, David D. 2009. *Needs Assessment*. New York: Oxford University Press.

Ruffolo, John, Paul Lee, and Duncan Stewart. 2010. Telecommunications Predictions 2010. http://www.deloitte.com/view/en_GX/global/industries/technology-media-telecommunications/tmt-predictions-2010/downloads

Smith, Shannon, Gail Salaway, and Judith Borreson Caruso. 2009. The ECAR Study of Undergraduate Students and Information Technology. http://www.educause.edu/Resources/TheECARStudyofUndergraduateStu/187215.

Snell, Steven. 2009. Mobile web design trends for 2009. *Smashing Magazine*. January 13. http://www.smashingmagazine.com/2009/01/13/mobile-web-design-trends-2009/.

Stanley, Tracey, Frances Norton, and Barry Dickson. 2003. Library project management in a collaborative web-based working environment. *New Review of Academic Librarianship* 9, no. 1 (December): 70-83.

Terriblyclever (Firm). Terriblyclever Design. http://www.terriblyclever.com/.

U.S. Department of Health & Human Services. Learn about usability testing—Test and refine. Usability.gov. http://www.usability.gov/refine/learnusa.html.

U.S. General Services Administration's Office of Citizen Services and Communications. USA.gov Mobile. http://mobile.usa.gov/.

Wamsley, Lori H. 2009. Controlling project chaos: Project management for library staff. *PNLA Quarterly* 73, no. 2 (Winter2009): 5-27.

Web Accessibility Initiative. 2008. Web content accessibility and mobile web. October 14. http://www.w3.org/WAI/mobile/.

———. 2009. Shared web experiences: Barriers common to mobile device users and people with disabilities. June 1. http://www.w3.org/WAI/mobile/experiences.

Weiss, Meredith. 2008. Results-based interaction design. *EDUCAUSE Quarterly* 31, no. 4 (December): 42-49.

Winston, Mark D., and Tara Hoffman. 2005. Project management in libraries. *Journal of Library Administration* 42, no. 1 (February): 51-61. doi:10.1300/J11v42n01-03.

Wiredu, Gamel O. 2007. User appropriation of mobile technologies: Motives, conditions and design properties. *Information and Organization* 17, no. 2: 110-129. doi:10.1016/j.infoandorg.2007.03.002.

Witkin, Belle Ruth. 1995. *Planning and conducting needs assessments: A practical guide.* Thousand Oaks CA: Sage Publications.

World Wide Web Consortium. 2008. Mobile web best practices 1.0. July 29. http://www.w3.org/TR/mobile-bp/.

Yandle. 2008. *Media chooser.* January 21. Flickr! http://www.flickr.com/photos/yandle/2208732061/.

Young, Jeffery. 2009. Teaching with technology face-off: iPhones vs. PC's. *Wired Campus.* February 25. http://chronicle.com/blogPost/Teaching-With-Technology/4547.